Grabbing Lightning

Grabbing Lightning

Building a Capability for Breakthrough Innovation

Gina C. O'Connor

Richard Leifer

Albert S. Paulson

Lois S. Peters

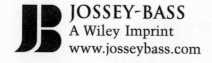
JOSSEY-BASS
A Wiley Imprint
www.josseybass.com

Published by Jossey-Bass
A Wiley Imprint
989 Market Street, San Francisco, CA 94103–1741—www.josseybass.com

Readers should be aware that Internet Web sites offered as citations and/or sources for further information may have changed or disappeared between the time this was written and when it is read.

Limit of Liability/Disclaimer of Warranty: While the publisher and author have used their best efforts in preparing this book, they make no representations or warranties with respect to the accuracy or completeness of the contents of this book and specifically disclaim any implied warranties of merchantability or fitness for a particular purpose. No warranty may be created or extended by sales representatives or written sales materials. The advice and strategies contained herein may not be suitable for your situation. You should consult with a professional where appropriate. Neither the publisher nor author shall be liable for any loss of profit or any other commercial damages, including but not limited to special, incidental, consequential, or other damages.

Jossey-Bass books and products are available through most bookstores. To contact Jossey-Bass directly call our Customer Care Department within the U.S. at 800–956–7739, outside the U.S. at 317–572–3986, or fax 317–572–4002.

Jossey-Bass also publishes its books in a variety of electronic formats. Some content that appears in print may not be available in electronic books.

Library of Congress Cataloging-in-Publication Data

Grabbing lightning: building a capability for breakthrough innovation/ Gina C. O'Connor . . . [et al.].—1st ed.

    p. cm.

Includes bibliographical references and index.

   ISBN 978-0-7879-9664-2

    1. Technological innovations—Management. 2. New products. I. O'Connor, Gina Colarelli.

  HD45.G64 2008

  658.4'062—dc22

2007045003

Printed in the United States of America

FIRST EDITION

HB Printing   10 9 8 7 6 5 4 3 2 1

Contents

Preface   ix

Introduction   xv

1   Management Systems for Innovation   1

2   Assessing Your Organization's Capacity for Breakthrough Innovation   23

3   The Discovery Competency   51

4   Incubation: The Long and Winding Road   81

5   Acceleration: Gathering Steam and Building Critical Mass   117

6   The DNA Innovation System   151

7   Incorporating the DNA: The Role of the Orchestrator   185

8   Getting Started: Initiating and Maturing an Innovation Management System   215

9   The Innovation Function   259

Appendix A: Companies Participating in the Breakthrough Innovation Research Study   275

Appendix B: Assessing Your Firm's Breakthrough Innovation Competency   279

Notes   305

Acknowledgments   315

About the Authors   319

Index   323

Preface

This book is about how to increase your company's capability for innovation—breakthrough innovation. You'll soon notice that we are innovation enthusiasts. Actually we're students of innovation, professors of innovation, innovators ourselves. We've studied large firms and start-ups, financial services organizations and manufacturing companies, public companies and private family-run companies. We've studied innovation in one form or another, as academics, for nearly two decades and worked with numerous companies beyond those we've studied to help them become more accomplished innovators.

Together, as part of a larger team, we have studied break-through innovation in large established companies, the subject of this book, since 1995, when the Research Program on Radical (or Breakthrough) Innovation at Rensselaer Polytechnic Institute's Lally School of Management and Technology originated. It has progressed in two phases: phase I ran from 1995 to 2000 and phase II from 2001 to 2006. In this Preface, we describe a bit about the research program so you'll understand how we've come to the conclusions that we present to you in this, our second book.

Our major conclusion is that innovation, which over the years has been a fad that comes and goes, is finally coming into its own. Innovation, we believe, is an emerging function in companies, much like marketing, quality, and finance have become functions in organizations over the past forty to fifty years. Companies are recognizing the need to develop an innovation capability that is sustained over time rather than a program du jour. They are struggling with how to do this because they don't know how. Nevertheless, they're focusing on the experiment and are investing heavily.

Our research program began in 1995 with a generous grant from the Sloan Foundation and the sponsorship of the Industrial Research Institute, a professional organization of R&D Managers, Directors and Chief Technology Officers of Fortune 1000 companies. The objective of phase I was to learn how breakthrough innovation (which we'll refer to as BI) projects were managed in large, established companies. We believed that management processes that worked well for incremental innovation would kill off breakthroughs before they got out of the starting gate, so we studied twelve projects in ten large, established companies over the course of five years. These were projects that senior leadership in the companies identified as potential breakthroughs. Over the five years that we tracked them, some were killed off, as expected, and others met with varying degrees of success. Several changed the game in their industries.

On the basis of that study, a team of us from the research program wrote *Radical Innovation: How Mature Firms Can Outsmart Upstarts*, in which we identified the challenges of managing under high levels of uncertainty on many dimensions (we highlighted four: technical uncertainty, market uncertainty, resource uncertainty, and organizational uncertainty).[1] The vast majority of projects we studied originated and progressed solely because of the strong will and persistence of a talented champion with ties to and protection from a senior management sponsor. Project teams and their leaders spent a great deal of time fighting against the norms of their companies, whose management systems, processes, and metrics were dedicated to efficiency, responding to market needs, and operational excellence. These were not the norms and infrastructure needed for the high-uncertainty environment of BI. We concluded that companies could do a much better job at developing management systems and infrastructures that supported, rather than antagonized, innovation champions and their teams. We coined a term, *radical innovation hub*, that we used to describe such a system. A hub is a central locus for BI activities: establishing, generating, and developing projects; providing coaching about dealing with ambiguity and uncertainty; accessing resources; networking; and gaining help for projects preparing to make the transition into business units or other receiving units.

Hub members would run interference between project teams and the mainstream organization and ensure that the connections with the mainstream organization occur as necessary so that the breakthrough is accepted into the company rather than rejected or ignored. The hub would develop the right kind of skills in people, oversee a portfolio of breakthrough projects, and set up appropriate governance mechanisms for each project and the portfolio. And it would play many other roles that we observed as gaps in companies' abilities to develop breakthroughs today. We noted the beginnings of a hub in two of the companies we studied, but that was all. And both of those were snuffed out by the end of the study period. Interesting, and discouraging.

As soon as the first book was published in 2000, we began phase II. We needed to know how companies that wanted a sustained BI capability were executing it. Was a hub the right approach? What alternatives were being tried? We went back to the IRI and proposed a second study, and the results of that study form the basis for this book. This book is about building an innovation capability that will last. It's about developing an innovation function. Companies are doing it. They're getting there. It's early yet, but it is happening.

For phase II we sought participation from companies that had a declared strategic intent to develop or evolve a sustainable BI capability. The majority of those that agreed to participate were just getting started. Twelve companies eventually were included in the study and participated for the complete observation period of three and a half years, from 2001 to 2004. We were amazed at the attention innovation capability development has gained since the late 1990s.

Our research team was a group of ten academics from three universities and several Ph.D. and M.B.A. students.[2] The team members represented had a wide variety of backgrounds, with expertise in technology policy, risk management, organizational development, marketing, information systems management, entrepreneurship, and strategy.

After an initial brief interview to understand their BI initiative—its history, mandate, and structure—we visited the company. We asked about all aspects of the management system for innovation. We met with the individual or board responsible for the BI capability development (we'll call this person or team the *hub leadership*). We met with the individual or team to whom the hub leader reported. We also met with the hub leader's staff and, in some cases, team leaders of projects that were part of the breakthrough project portfolio.

Every six months for three years following our site visit, we held a follow-up interview with the hub leader and whomever else she or he considered important for us to speak with at the time. In all cases except one, we spoke with the company president or a senior vice president or a chief technology or chief strategy officer of the company.

About six months after we began our interviews, we began receiving inquiries from other companies that had heard of the study and wanted to participate. We formed a validation group comprising these firms that met at RPI every six months for the duration of the study. Those meetings were structured to address topics that were emerging as important in our interviews. Each company in the validation group (there were nine companies in all) reported on its practices, and we'd pose hypotheses for debate in those meetings. They were extremely helpful in providing even broader and richer context to what we were seeing in our initial twelve companies. Table P.1 lists the companies that provided projects for us to study in phase I, the twelve initial companies that volunteered to participate in this second phase, and the

Table P.1 Study Participants

Phase I (1995–2000)	Phase II Initial Sample (2001–2005)	Phase II Validation Sample (2002–2005)
Air Products	3M	Bose
Analog Devices	Air Products	Dow-Corning
DuPont	Albany International	Guidant
GE	Corning	Hewlett-Packard
GM	DuPont	Intel
IBM	GE	P&G
Nortel Networks	IBM	PPG
Polaroid	Johnson & Johnson (Consumer Products Co.)	Rohm & Haas
Texas Instruments	Kodak	Xerox
United Technologies (Otis Elevator Division)	Mead-Westvaco	
	Sealed Air Corporation	
	Shell Chemicals	

validation set of companies for phase II as well. Each company is briefly described in Appendix A.

All of our interviews were taped, and the stack of transcripts stands three feet tall. Our research team met regularly to debate what we were learning. We had meetings that were facilitated by our IRI subcommittee cochairs, we had meetings in restaurants, we had meetings in conference rooms—all accompanied by laptops, flip charts, and Excel spreadsheets. The meetings were challenging, exhausting, and fun. Some of our views emerged in those large meetings and others through quiet study and smaller discussions. We've vetted our findings with our IRI sponsors, with members of the Institute of Electrical and Electronics Engineers (IEEE), the Product Development and Management Association (PDMA), and the Strategy Executives Group of the Conference Board.

Midway through our data collection period and again at the end, we held workshops in which we invited members from all participating companies to discuss and validate what we thought we were learning. These too were challenging sessions. We delivered our findings, and we were questioned. We continued to develop our thinking. We watched their discussions among themselves, and we continued to learn. Even today we feel free to call on them for clarification of issues we're struggling with as we continue to read for clues in those interview transcripts. They are a wonderful set of participants, they're all interested in furthering their own learning and that of the collective and larger innovation community, and we are honored to have had the opportunity to work with them.

January 2008
Troy, New York

Gina C. O'Connor
Richard Leifer
Albert S. Paulson
Lois S. Peters

Introduction

Leaders of established firms have worried about rejuvenation and new business creation for a long time. The more firmly established a company becomes, the more concerned with growth and renewal its leaders are. It's easy to depend on the same product lines, the same customer segments, the familiar business models because they're well understood. But the rest of the world does not stand still. Technology and competition march on, and companies that are not ahead of the pack by a long shot in developing new paths to growth by creating new markets and breakthrough innovations face commoditization, marginalization, and relentless competition for smaller and smaller market share. Companies need to develop a capability for breakthrough innovation. It raises the chance of long-term success.

Besides, innovation can be fun. Really fun. And sometimes painful, but still fun. We find that many people just can't help themselves. They are driven to participate in innovation-related activities despite the associated risks. Innovation brings meaning to their work lives. They want to be involved in nurturing new businesses and bringing leaps of value to the market. They're proud of their companies and want to believe their companies are capable of offering much more.

We know of stories of breakthrough innovation (BI): Corning's development of optical fiber, Motorola's pager device, GE's CT scanner. But these occur infrequently, irregularly, and unpredictably. Companies need to develop a capability for BI so that it occurs over and over again rather than by happenstance. Some of them already have this capability.

Innovation is emerging as an organizational function just like marketing has become established as a function in companies today. Forty years ago, few companies had marketing departments. None had chief marketing officers. Today many do because company leaders came to realize that marketing-related activities (selling, public relations, competitive intelligence, branding, market research) needed to be ongoing and ever present. Marketing could not be a program du jour or a special budget request. It needed to be embedded in the ongoing activities of the organization. The same thing has occurred with quality assurance, operations, and information systems.

So it will be for innovation. BI isn't a once-in-awhile sort of thing. Although large, mature companies have experienced it in this way, more can be done. Certainly innovation is vulnerable to the ups and downs of the business cycle, the industry cycle, and even the company's own cycles, but it should not be extinguished. It needs to be continually present, just as marketing is continually present, even though the marketing budget may be squeezed or amplified given the company's condition at any point in time. We need to develop the innovation function in companies, and by this, we mean BI.

Companies have experimented with different approaches to developing a BI capability over the past forty to fifty years. None of these approaches, however, has addressed innovation as an ongoing business function that requires career paths, appropriate metrics, a clear mandate, and all the other elements that a sustainable function requires.

Historical Approaches to Innovation in Established Companies

In 1985, Gifford Pinchot published *Intrapreneuring: Why You Don't Have to Leave the Corporation to Become an Entrepreneur*, in which he prescribed that corporate entrepreneurs learn to seek out senior sponsors for protection, break rules as needed, and get the new business created within the firm in spite of the resistance they'd face.[1] Intrapreneuring became a mantra to those in companies with new-business-creation ambitions. But Pinchot's model is one of exception rather than one of organizational capability. It depends on champions, on exceptional individuals. And we know that there are many breakthrough ideas that lie fallow because the idea generator and the opportunity recognizer aren't always endowed with champion-like or maverick-like personalities.[2]

A skunk-works approach is another model companies have tried. This approach originated in the Lockheed Corporation during World War II and was used successfully to develop a high surveillance airplane that was used in the 1950s.[3] Skunk works are relatively small work groups, physically separated from mainstream operations so as not to be subjected to the burdensome rules, overhead, or bureaucratic pressures of the corporation. They're supposed to be sequestered to let the creative juices flow. IBM's PC junior was developed in this manner, as were DuPont's development of Lycra and Raychem's development of the microwave oven. Japanese companies such as Cannon and Honda have also used this approach effectively. In all of these cases, the skunk-works approach was designed to develop a particular product and was not used as a systematic process for developing a pipeline of innovations. The famed Xerox PARC, designed as an ongoing skunk-works group to develop such a portfolio, failed as a source of growth and renewal for its mother company. PARC's businesses were adopted and commercialized by others but rejected by its own sponsoring company, Xerox. PARC was criticized for being too isolated from the company.

Another model companies have tried has been a corporate venturing unit, which emerged in the late 1960s and early 1970s, inspired by the successes of venture capitalists themselves, who had backed start-ups such as Digital Equipment Corporation and Raychem.[4] These were venture funds that companies established to invest in start-up organizations they perceived to have potential for high growth, expecting to capitalize financially, just as stand-alone venture capital funds do, but also to strategically help the development of new technologies and markets from whom the firm would benefit. More than 25 percent of Fortune 500 firms had corporate venture investment programs during those years. When the capital markets declined in the late 1970s, corporate venture capital funds suffered the same fate as stand-alone venture capital funding. In the early 1980s, venture capital money rebounded, and corporations reinitiated corporate venturing activities, accounting for 12 percent of venture capital investing by 1986, only to be severely diminished once again after the 1987 market downturn.[5] Almost 40 percent of corporate venture programs were abandoned within four years of their initiation, and by 1992, corporate venture capital investments accounted for only 5 percent of total venture capital investing.[6] A third wave of corporate venture capital activity occurred in the late 1990s, again following the boom of venture capital activity that was fueled by Internet and other technologically based opportunities.

Between 1995 and 1999, the number of U.S. companies that made corporate venture investments, looking for new revenue sources through the commercialization of new technologies, increased from 62 to 415, with investment dollars exploding from $542 million in 1996 to $16.5 billion in 2000.[7] The business venture groups at Lucent Technologies, Cisco, and Nortel Networks were followed by many students of the corporate venture capital model, as they had healthy portfolios. These efforts, however, still suffered from lack of persistent experience and expertise in the managerial strategies necessary for operating in the regimes of high uncertainty that BI demands. In addition, although the portfolios were healthy from a financial standpoint in many cases, they failed to contribute to the overall strategic growth and renewal of the firm.[8] Many of the portfolio companies never contributed to any new business growth of the mainstream sponsoring organization.

At the same time that companies were placing investments in external small companies or internal ones that they spun out, a wave of internal corporate venturing experiments was taking place. During the late 1970s and 1980s, new ventures divisions were forming to incubate, inside the organization, projects that didn't fit the current business models or organizational structures of the firm. Most, however, were eventually disbanded, since the businesses were not achieving large-scale growth quickly enough, and managerial patience wore thin.[9]

So companies have tried many approaches in recent history, but none of these models has resulted in a sustained successful BI capability. Why? Because they're dependent on individual champions, because they're not part of the company's strategy or strategic intent, or because they were singularly focused on one project (IBM's move into the PC business, for example). IBM didn't know, as a company, how to enact new business creation any better following that successful experience than it did before. There was little learning and little impact on the functioning of the organization largely because it operated in isolation from the rest of the company. How was the learning captured and put into repeated practice?

Treating BI as a temporary program, an appendage to the mainstream organization, or a one-off project that the company invests in on an ad hoc basis is risky. It makes innovation vulnerable to the whims or fancy of whoever is in charge, the current financial picture of the firm, or the condition of the capital markets outside the firm. It's either on or off, go or no go.

Nortel Networks, for example, established a business ventures group in late 1996 whose purpose was to capture the creative ideas of those in the company with entrepreneurial talent and drive and develop those opportunities into businesses that would either be spun in to the company or spun out as stand-alone ventures. Joanne Hyland, vice president of the group, worked hard to find and develop a team of people who could coach the fledgling startups, help with market and economic analyses, and help negotiate external financing deals. She developed processes for finding and vetting the opportunities to develop a portfolio, set up oversight boards for each of her portfolio businesses, assembled a governance team to help manage the portfolio itself, and developed processes that helped the entrepreneurs manage their way through the natural chaos of BI. Over the course of the next three years, the small group began to experience success. Of course, not all of the ventures succeeded; such is the nature of innovation on the edge. But a number of the ventures were successfully spun out of Nortel, and others were moved into the business units. Then the CEO retired and was replaced by a person who preferred acquisitions to internal organic growth. Within a year, the business ventures group was shut down, and all of the learning that had been built up in Hyland's organization was lost. The management system she built was disbanded. All of those in the group left, and with them, out walked the organizational memory for how to create new businesses. If Nortel wanted to grow organically today, it could not draw on the experiences of the business ventures group because no one from that group is there anymore.

Similarly, 3M's chief technology officer in 2001, Paul Guehler, wanted to resurrect 3M's new ventures group, which had been completely defunded fifteen years ago. The organizational memory, the practices, the people, the pipeline of projects were gone. He'd have to start from scratch.

So over the past thirty or forty years, companies have invested time, people, and money into building new business creation entities, structures and processes, only to shut them down. What a waste. We not only need to find ways to maintain a constancy of the innovation function itself, but also to learn how to power it up and down as needed.

Misunderstandings About Breakthrough Innovation

In our discussions and interactions with companies, we have found a number of misunderstandings or gaps in understanding of BI, which lead to confusion about how to manage it.

Lack of an Innovation Strategy

Most companies have clear strategies for operational excellence and even new product development, but very few have explicit strategies related to BI. It's as if they believe breakthroughs will happen serendipitously or, even better, as the result of leadership's declaration that they're needed . . . now! This is probably exemplified by the initial resistance to and difficulty that established companies had in making total quality management a formal company strategy. Indeed, it took heroic efforts by W. Edwards Deming, Joseph Juran, and Philip Crosby, among others, to hammer away at U.S. companies regarding the importance of recognizing customer requirements for high-quality products. If it weren't for the Japanese automobiles taking substantial market share away from Ford, Chrysler, and GM and almost putting those companies out of business, this new approach to doing business probably would not have been implemented. Indeed, it still took ten years, until the early 1990s, for the U.S. automobile makers to finally understand this concept. Will it take that level of urgency, a life-threatening series of events, for companies to recognize the importance of developing an innovation capability?

Our contention is that BI will not happen as an organizational capability unless it becomes part of the strategic portfolio of the company on a par with operational excellence and new product development. This means senior management must consider BI a corporate priority, be visible and audible with respect to supporting it, and provide corporate resources to get it done. As Nick Donofrio, IBM's executive for innovation and technology (he used to be called simply IBM's chief technology officer), said, "In the 21st century, innovation is my job. It is the most important thing I do for my organization. What I need to do is to take my organization away from where it is now and move it to a place of higher value."[10]

Making BI a strategic priority protects it from becoming a flavor-of-the-month program without long-term survivability.

The natural corporate antibodies of resistance to new ideas and programs will overwhelm nascent BI activities unless they are protected and given organizational validity, support, and attention.

Lack of Distinction Between Invention and Innovation

An *invention* is a new object, process, or technique. Firms that invest heavily in R&D reap the benefits of invention: many patents and many published papers. But that's where it ends unless they also know how to innovate. *Innovation* is the introduction of something new to the marketplace. All of the rest of the work to develop an invention into a marketable offering and a business platform comprises innovation. Inventions are the inputs to innovation. Inventions, on their own, need not ever come to market. There's much work involved in innovation beyond the initial discovery. We call that work *incubation and acceleration*, and we'll introduce you to those, along with discovery, as the three major building blocks of a BI capability.

For many companies, placing a strategic emphasis on BI will result in simply directing increasing amounts of resources toward it. After all, companies reason, if it's important, they should invest in it. We would therefore expect that increasing investments in R&D would lead to increasing amounts of innovation, which should then lead to increasing organizational success. This line of reasoning led us to investigate these assertions carefully. We examined ten years of data relating R&D expenditures as a percentage of sales to sales growth, profitability, total stockholder returns, and degree of innovativeness in twenty-two industries for about twelve companies in each industry. For innovativeness, we used a portion of the *Fortune* magazine annual Most Admired Companies survey that asked industry peers (executives, managers, consultants in an industry) to rate companies in their industry. Each industry ranking is the result of the opinions of roughly five hundred people in that industry.

What we found was quite surprising. First, we found no relationship between R&D expenditures as a percentage of sales and innovativeness. This means that simply investing in R&D more won't lead to more innovation. Second, we found that R&D expenditures as a percentage of sales was consistently unrelated to sales growth and margins or stockholder returns, and in some cases, it was even negatively related to stockholder returns. This means that R&D investments don't, in and of themselves, lead to better company performance. Third, we found that innovativeness was not related to profitability, but innovativeness was related to sales growth and stockholder returns. However, for firms in the sample that were rated as highly innovative, these relationships did not occur until after the third year. This means that it takes at least three years for innovation activity to have a financial impact on the company.

These findings underline the importance of innovation to long-term company success. More important, they demonstrate the importance of investing in incubation and acceleration infrastructures and management systems rather than investing more in R&D. Although investment in R&D is obviously necessary, more than the industry average will not yield a return on that investment unless there is a concomitant investment in the rest of the innovation management system.[11]

Inappropriate Expectations for Breakthrough Innovation

Even when senior management recognizes the importance of BI, there is often a lack of understanding of the investment required in terms of time horizons, complexity of creating new markets and new businesses, and the nature of the returns. The short-term results are palpable, but they're not measured in financial or stock return results. Immediate results include an energized workforce, renewed optimism, imaginative problem solving, new networks, and new opportunities. After a few early successes, there's a buildup of confidence, a 'can-do' mentality. But it takes a long time for a portfolio of highly uncertain opportunities to be nurtured along and yield commercial and financial success. Typically management measures success in that way, and so traditionally, breakthrough innovation programs don't last. That's why we need to develop an innovation function that has appropriate metrics and evaluation criteria.

Managing Breakthrough Innovation

Applying what we already know about managing ongoing operations to managing business innovation would be a big mistake, but it's done all the time because it's what we know how to do. But managing BI is fundamentally different from managing other organizational activities, even new product development. Table I.1 highlights these differences.

Our objective in writing this book is to help you address the problems of implementing an organizational innovation capability. It's based on what others have done so that you can build on their experiences. We are not advocating that you change your ongoing operations. We are advocating that you add a new capability . . . : a BI capability. As Table I.1 shows, it's quite distinct from what the mainstream organization does, and so the innovation function requires a management system that works for it rather than against it. It's time that innovation was not an exception but rather a part of every company.

Throughout this book, we point out the difficulties and challenges companies we've studied have faced as they are developing this BI capability. We also point to their successes, and there are many. No one has it all down yet. We share these stories so you can learn from the experiences of others and move more quickly to getting innovation institutionalized in your organization.

We lay out a plan and strategy for developing this function given where you are now and your strategic intent. Not every business will compete on the basis of BI, but every company should

Table I.1 Differences Between Managing Ongoing Operations and Breakthrough Innovation

Management Issue	Ongoing Operations	Innovation Function
Innovation objective	New products to extend existing businesses	Breakthrough innovation based on advanced technologies and business models yielding platforms to create new businesses
Time orientation	Present and near future (one to two years)	Far future (three or more years)
Strategic objectives	Efficiently and effectively satisfy and even delight current customers; plan for next generations	Create new markets based on new-to-the-world performance features or order-of-magnitude improvement in known features or cost; develop strategic intent for domain focus and allow opportunism
Culture	Operational excellence: customer intimacy and execution skills	Cultivation; employee intimacy and new business creation skills
Risk profile	Risk averse with a focus on system efficiency	Risk mitigation through staged learning
Opportunity selection	Customer driven, based on voice of the customer market research	Vision and possibilities tied to strategic intent
Investment timing and revenue focus	New products in six to eighteen months; profit-and-loss management with in-year revenue streams	New businesses in three to five or more years; return on investment over long term with portfolio management to hedge bets
Project management processes	Phase gate and concurrent engineering	Discovery-driven processes and learning plans

have such a capability in place. The focus of the book is to understand the organizational system that's needed to sustain innovation in your firm. We do this in parts. The first seven chapters describe the innovation system.

Chapter One describes what a management system is and details how and why the management system for innovation must differ from one that supports a culture of operational excellence, the dominant culture in most established companies. Then we introduce the building blocks that comprise an innovation function. We refer to this as the DNA genetic structure of BI (Discovery, iNcubation, and Acceleration—note that we've cheated a bit on the "N") and make the case for entrenching the innovation function right into the company rather than as an appendage.

Chapter Two introduces the idea of an organization's capacity for innovation. Why can some companies do this more easily than others, and why can some companies be more open to developing BI capability now rather than several years ago or several years hence? When capacity is rich, what do you do? When it's strained, what do you do?

Chapter Three is all about the discovery competency. Discovery is not the same thing as R&D. What activities are involved? What are the implications for R&D? How do the innovation management system elements play out in the discovery system? What challenges are firms facing in discovery, and what are some keys to doing it well?

Chapter Four examines the incubation competency, the most difficult and overlooked of the three BI building blocks. If companies can learn how to master incubation, they greatly expand their options for new business creation.

Chapter Five focuses on the third building block, acceleration. Acceleration was a surprise to us in the research program. Building critical mass in a new business must be handled carefully, and this chapter addresses the activities involved, approaches companies are taking to acceleration, and pitfalls associated with not attending to it properly.

The discovery-incubation-acceleration system is the subject of Chapter Six. We find that few companies have a balanced effort across D, N, and A. What are the implications of this? What are the DNA system-level activities that have to be handled? What are some of the organizational structures that firms have developed to ensure that they are all happening? Finally, what are appropriate metrics for a DNA system?

Chapter Seven is about orchestration. Let's suppose that you've built an elaborate DNA system and it's producing breakthrough businesses, but the mainstream organization cannot absorb your innovations. What has gone wrong? Who is the orchestrator, and what are some organizations' approaches to orchestration? How do you orchestrate in munificent times? In lean times? What are some keys to effective orchestration?

In Chapter Eight, we address how to get started developing a BI capability. What can you expect to face as you initiate this in your company? If you've already been at it awhile, what challenges have we seen companies encounter that you might avoid? It turns out there are some fairly predictable challenges that you may face, and we discuss options for how to handle them.

Finally, Chapter Nine describes our thoughts on how to make innovation a sustainable activity. What does a mature BI capability look like? What does it need in companies in order to flourish?

At the end of most of the chapters is a set of questions to help assess your company's BI competency. Go through and score each question twice: first for how competent you perceive your company is now, and then give a score that you wish represented your company. List your average scores for the set of questions in each chapter in Appendix B at the back of the book and then plot them on the graph in the appendix to see your company's BI profile.

It's time to recognize innovation as an organizational function and get started developing it in your company.

Grabbing Lightning

1

MANAGEMENT SYSTEMS FOR INNOVATION

Companies are clamoring for innovation—breakthrough innovation. Every executive we've spoken with in the past ten years wants it. The trouble is they don't know how to make it happen. "We've atrophied," a senior technical leader at GE's global research center told us four years ago. "We've been so focused on responding to our business units' immediate needs and on taking cost out of our products that we've forgotten how to turn advanced technologies into breakthrough products and businesses."

Struggling to Market

There are many reasons that companies struggle with getting breakthroughs to market. A few examples from project teams that we studied in the first phase of our research in the late 1990s illustrate some of the problems companies encounter all too often. Ultimately all of these companies developed a complete breakthrough innovation capability so they could overcome the sorts of challenges presented here.

Air Products, Then and Now

The management of Air Products, a large industrial gasses and performance materials company, knew they had to get growth through new initiatives. In the early 1980s, they realized they'd missed a major technological shift in air separation processes—methods to distill air into purer levels of its component parts, such as oxygen,

1

nitrogen, hydrogen, and other gasses—and had been left behind. They hired a number of scientists with different backgrounds to help diversify their technological prowess in the mid-1980s. Part of that group was a small band of material scientists with expertise in ceramics materials, which Air Products' technical community believed could potentially offer new-to-the-world features in gas and air separation processes that the company was famous for. But the investment yielded little, and the group was disbanded in the early 1990s.

Several of these scientists, however, had come up with an idea for gas separation that was technically challenging but looked as if it could be a breakthrough in making high-purity, high-flux oxygen. The process would disrupt the oxygen distribution business because the oxygen would be made on site and would not need to be transported in the cylinders that are used today. These scientists labored diligently and developed Gantt charts, used stage-gate processes, and developed economic models. They began to realize that some of those project management processes could not work, since most of the markets they needed to contact were not ones in which the company currently participated, so they added a market analyst who came over from the field sales force. But then many market opportunities that they had initially explored had dried up. Still they labored on, now partnering with another development firm and securing government funding. Then the team's leadership was changed. They were questioned by the corporate executive board regarding their slow progress, but since much of their funding was external, they were not a heavy drag on the firm and so have been allowed to continue. Thirteen years later, they have working prototypes and some first customers.

Today Air Products has a functioning commercial development office (CDO) with staff who help develop such opportunities. They're connected to the company's newly formed growth board, so that the opportunities are aligned with company strategy for the future; in fact, they influence that strategic growth plan. They reach into various parts of the organization for the help they need, be it R&D support, market connections, or manufacturing. The oxygen generation project continues, but in the meantime, the CDO has vetted and nurtured numerous other business opportunities. With the help of a staff who understands how to probe into new markets, recognize opportunities for the technological prowess the company has, and leverage networks inside and out of the company, a portfolio of breakthroughs, rather than just one, is on the horizon. Air Products is developing a breakthrough innovation capability.

IBM, Then and Now

In the early 1990s, Bernie Meyerson, an IBM research fellow, happened on a discovery, an alloy of silicon and germanium, that he believed could become the basis for high-performance new transistors with switching speeds up to four times faster than those of traditional semiconductors, with applications in the exploding wireless communications market.[1] An important benefit was their ability to operate using only a fraction of normal power requirements for competing technologies. IBM's ability to mass-produce silicon germanium would make it possible for hardware manufacturers to substitute chips made from this material for more costly, power-hungry, and exotic alternatives, such as gallium arsenide. Silicon germanium semiconductor technology offered a breakthrough price-to-performance ratio not available from these existing component technologies. Best of all, the new chip material could be manufactured with the same costly fabricating equipment used to make conventional silicon chips, potentially avoiding billions in new capital investments. The problem was that IBM was not interested. The company had made its money selling mainframes, not chips. And although it made chips, those were used only for internal customers; they were never sold to external original equipment manufacturers.

But Meyerson could see the market coming, and he knew IBM had to be there. For most of the months and years that they worked on the development of silicon germanium chips, he and his ad hoc team of collaborators had operated as a band of mavericks—tolerated but not officially sanctioned by the IBM R&D establishment. They made progress using bootlegged time and resources when the technology was outside IBM's strategic framework. Meyerson's relationship with Paul Horn, the senior vice president of R&D, was one of mutual trust and respect. Horn allowed Meyerson to continue with his work, though without his official approval. The silicon germanium project was broadly viewed as an irritating virus within the R&D host, which used the usual organizational antibodies to neutralize it: withholding funding, general naysaying, and subtle signals that it might not be "career smart" to associate with the project. Meyerson was immune to the implication of these signals due to his status as an IBM fellow. His project continued.

Meyerson went to great lengths in preparing his presentations, showing his experimental data and contrasting them to performance data for pure silicon and for the leading contender for higher-performing, next-generation chips: gallium arsenide. Although the project was essentially a low-priority project among senior leaders within his own company at the time, Meyerson presented his data at a number of professional conferences and published articles in scholarly journals. Conference attendees representing Northern Telecom, Analog Devices, Hughes Electronics, and other leading companies recognized the potential of silicon germanium research and approached him to express their enthusiasm. These fellow scientists could see important applications of silicon germanium technology in their businesses, particularly in telecommunications. These potential uses were a revelation to the IBM researcher and helped him to fill in the details of what had initially been a fuzzy vision.

When CEO Lou Gerstner dramatically shifted corporate strategy to include the sale of chips to external customers, Meyerson's project quickly gained legitimacy. Now it was legitimized, but he envisioned a different sort of business model than IBM was used to. The company had a long history as a volume producer of memory chips, an arrangement in which other firms provided the higher-value application design functions. Meyerson was aware of that history and knew that acceptance and funding of his project would probably lead to a similar outcome: IBM would get the high-volume chip-fabricating part of the value chain, while someone else would handle the smaller but higher-profit business of design. Meyerson believed that IBM should go after more of the value chain associated with the silicon germanium chip and expand IBM's chip-making activities beyond its traditional foundry operations. He and the project's first business manager pushed for broader involvement in the value chain and resolved to work only with companies that were willing to let IBM in on the chip design issues relevant to the application. Meyerson hoped to learn these skills and eventually co-opt that part of the business for his company. He wanted IBM to use the development of the new chips as an opportunity to expand its design capabilities. In the end, his goal was partially accomplished when IBM's CommQuest subsidiary was slated to design silicon germanium–enhanced chip sets for next-generation cellular phones. Other efforts were ongoing, like hiring a workforce trained in the specialty design coding work that was needed.

The silicon germanium chip was announced in October 1998 and eventually became an industry standard, but it had to be rescued many times. Its success can be attributed to Myerson's technical brilliance, rebellious personality, and tight connection to the senior vice president of research, Paul Horn. Meyerson is a one-man show. He refused to give up on his idea even though it contradicted the accepted technical wisdom, fell outside the strategic boundaries of the firm at the time, and met with substantial organizational resistance. His battles to get his idea heard ended up with—in his words—"lots of blood on the walls," but his passion and perseverance eventually won out due to his personality and position. However, it could not be duplicated and could not be a model for a system design for repeated innovation.

Beginning in 1999, a systematic approach to innovation was initiated, and today, IBM has a system in place to provide support for people like Meyerson: the emerging business opportunity management system. Entrepreneurial scientists or novel ideas that crop up anywhere in the company now have a place to go to get management attention, resources, and help. Scientists make great discoveries, but few of them indeed are also willing and able to pursue customers, negotiate alliance deals, develop a business model, and consider plant location decisions, as Meyerson did, all the while working to convince the company to adopt the initiative as part of their strategy for future growth.

The emerging business opportunity system is not dependent on a single senior leader. It encompasses a team with the responsibility for breakthrough innovation (BI) within IBM. This team also has a staff group of consultants who can help guide teams like Meyerson's, so that they are not left to handle all aspects of innovation on their own.

IBM and other large companies have always had characters like Meyerson. They're finally realizing through the school of missed opportunities that they need a more systemic approach to enable breakthrough innovation.

Building a Capability for Breakthrough Innovation

Large, established companies have never excelled at breakthrough innovation. Their management systems are designed to ensure highly reliable, repeatable processes. Everyone sticks to a plan based on market research and competitive trend analyses that tell them what to do. Their strategies are driven by financial objectives that they promise Wall Street. For the most part, the objectives are short term in nature because much of executive compensation is tied to quarterly performance.

Breakthroughs crop up once in awhile in big companies, but they occur because impassioned champions don't quit. They break the rules and find protection from one or two powerful senior leaders who believe in them. Why must this be the case? Large, established companies have access to the money, brains, and market power that they can draw on to make things happen. It's a waste to allow BI to happen by chance alone. And senior leadership knows it. They also know they can't save their way to the future or acquire companies as their only growth strategy. Research shows that firms that successfully commercialize breakthrough innovation reap above-normal returns and higher-than-average market value over the long term, but they must have adequate systems for managing high uncertainty in place to help leverage their investments.[2]

Since we wrote our first book, we have been amazed at the extent to which companies are investing in developing systems for breakthrough innovation.[3] They are experimenting with building innovation functions, departments, and disciplines that serve the objectives of breakthrough innovation; beginning to think of portfolios of breakthroughs; worrying about how to professionalize the career path for those involved in breakthrough innovation; and concerned with governance models. In short, they're developing management systems for innovation that parallel the finely tuned management systems for operational excellence.

Corning, for example, now has a vice president for strategic growth and new business development, a newly created position, who reports to the chief technology officer and operates out of the research and development division. His initial mandate was to find and articulate opportunities for breakthrough innovation. He has a team of people, organized as the Exploratory Marketing and Technologies Group, whose charge mirrors that of those in Exploratory Research: find new opportunities. Over the past three years, Mark, the vice president, and his team recognized that finding opportunities wasn't enough, so he built a business development team whose members work with the breakthrough project teams to nurture them through the development process. Why? Why can't R&D do that? Because there are business questions that arise:

- How do we build this as a new business?
- What applications might there be?
- Who should we partner with?
- How do we go about building a new customer base?

- What are the economics?
- Where should this reside within Corning?
- What path should the technology development take?

These questions aren't new; other writers have described these in previous books.[4] What is new is that Corning is doing something about this in a systematic way. It has a dedicated team of people who are developing this expertise, working with a portfolio of projects to help develop the business part of the innovations. So is IBM. The senior vice president of strategy, Bruce Harreld, has been leading a charge since 2000 to develop and run a management system for horizon 3 (H3) businesses, IBM's label for far future opportunities, that is, breakthrough opportunities. Everything about an H3 is different from a horizon 1 (H1) business, which is a mainstream business. Performance is measured differently, and people are recruited and then reviewed differently. Everyone in general management at IBM has had to undergo training to understand these differences.

At Sealed Air, CEO Bill Hickey and the vice president of engineering are building their infrastructure, which they know they need. Who's in charge of finding new ideas? Who builds them out into new businesses? What kind of oversight has to happen to make sure the new ideas do not get squelched once they move to the operating units? Sealed Air recently installed a corporate business development function that nurtures these opportunities. They're learning how as they go along, but they're dedicated to making it happen. Sealed Air's Business Innovation Board oversees the BI portfolio. They link the portfolio to the strategic intent of the company—not their current strategy, but their vision of the company's future businesses.

At Kodak, a company that has been faced with profound technological change in its core businesses, the Systems Concept Center (SCC) was the BI hub for more than ten years (1994–2004). Those in the SCC found that business ideas that were generated and tested could not survive when they moved into operating units, so they built an accelerator: an identified group with its own performance metrics, people, and oversight board who nurtured projects in a high-growth phase. The reason this group came about is that nothing else had worked for Kodak. Mainstream businesses could not find the talent, the interest, the understanding, or the resources to invest in small, high-potential breakthrough businesses that weren't to the level of predictable sales and cost levels. It wasn't their mandate and wasn't part of their performance requirements, so it didn't work.

The interesting thing about all of these companies is that none of them is following a skunk-works model. They're not relying solely on placing venture capital bets in external companies. They're not relying on champions to get it done against all odds. All of them are experimenting with making BI happen as part of the heart and soul of their company rather than as an offshoot, afterthought, or secret. They are expecting to simultaneously deliver on today's pressures and the future's uncertainties. "Senior leadership realized that while our responsibility is the long-run health of the organization, we spent most of our time on immediate problems," Bruce Harreld at IBM told us. "We had to change that."

Companies are experimenting and are trying to figure this all out. We have been studying twelve of them in depth for four years and another nine who check in with us regularly because they too want to learn. All of these companies qualified for participation in our study because they have a declared strategic intent to develop a breakthrough innovation capability (BIC)—not the capability to allow a maverick to sneak through the system but a management system whose objective is to enable breakthroughs over and over. Not one of these companies, in our estimation or their own, has a complete high-functioning system yet.

Our opportunity has been to observe them, see their struggles and their victories, and identify pieces of systems in each of these companies that, taken together, can create a BIC. From these insights we've developed a model that we have been using and testing successfully. It provides the building blocks and modus operandi for developing innovations that will change the game, shake up industries, and shake up companies' existing businesses by bringing dramatic levels of new value to the marketplace. These are the risky, uncertain investments that make company executives who need to respond to shareholders and Wall Street analysts very nervous. It takes courage, but also discipline and persistence stemming from an innovation management system. To become an integral part of any company on an equal footing with operations, marketing, finance, human resources, quality and other core functions, innovation, we believe, needs to become its own discipline, its own function in companies today. And we see the signs. It's early yet, but the cues are apparent.

This book describes an integrated management system for breakthrough innovation, details how it must function and how it interacts with the mainstream organization. It also highlights challenges that firms face as they develop these systems, which, in many ways, require managing in counterintuitive ways and helps set companies on the path toward developing that capability.

A Primer on Breakthrough Innovation Management Systems

In this next section we offer some definitions and frameworks to guide the discussion throughout the rest of the book. We start by defining breakthrough innovation and then go on to the elements of a management system for innovation. Finally we define what a breakthrough innovation capability is. Once we have those concepts in place, we can develop them more completely in the chapters that follow.

What We Mean by Breakthrough Innovation

There are many definitions of *breakthrough innovation*. Some believe breakthroughs disrupt currently functioning markets,[5] while others define a breakthrough as anything that earns the firm a standing in a new market domain. Some say breakthroughs are predicated on new scientific discoveries,[6] while others suggest that many breakthroughs are the result of clever business model innovations.[7] Indeed, there are any number of ways to achieve breakthrough innovation.

Working with our set of twenty-one companies as well as the Research on Research subcommittee of the IRI devoted to this project, we defined a breakthrough innovation as the creation of a new platform or business domain that has high impact on current or new markets in terms of offering wholly new benefits *and* high impact on the firm through expansion into new market and technology domains, increased revenues, and ultimately increased profits.[8] These high-impact levels, though, come with a set of challenges and dilemmas. As the long development periods unfold, risk and uncertainty associated with breakthrough opportunities abound. Companies find themselves in the situation of having to develop new technological competencies, create new market spaces, acquire different resources, and adapt their organizational structure to house the resulting new business, because breakthroughs may not easily fit into the current structure of the company.

The firms in our study use different terminology for breakthrough innovation. We've heard terms like *scope change innovations*, *game changers*, *moonshots*, *radical innovations*, and *rockets* to communicate the idea of breakthroughs. In fact, most of them told us they consider several different levels of innovativeness. We list some of their classifications in Table 1.1. Your company probably has one as well.

Some of these categorization schemes are organized by time horizon, some by fit with the current organization, some by degree of technological advance, some by impact on the market. But all of our company participants agreed that the categories in the far right column, no matter what they were called, were riskier and more uncertain than the ones listed on the left or in the center.

So, no matter what definition we use for *breakthrough innovation*, we are dealing with innovation opportunities that offer the promise of new growth platforms. They also may take the company into technology domains, business arenas, and domains of expertise that are unfamiliar and, in fact, may not yet even exist.

Table 1.1 Innovation Categories

	Degree of Uncertainty and Ambiguity		
	Relatively Low	Moderate	High
Company 1	Horizon 1	Horizon 2	Horizon 3
Company 2	Making the most of what we have	Getting new business	Breaking new ground
Company 3	Incremental	Platform	Breakthrough
Company 4	Today	Tomorrow	Beyond
Company 5	Incremental	Major Improvements	Step-outs
Company 6	Incremental	Substantial	Transformational
Company 7	Business unit projects	CEO projects	Advanced technology programs
Company 8	Incremental	Longer term	"We don't have a clue"
Company 9	Aligned	White space projects	Gray space (multialigned)
Your Company			

Up to this point we've discussed breakthrough innovation and its various incarnations. But this book is about how to build a breakthrough innovation capability. A *breakthrough innovation capability* is the ability for a firm to commercialize breakthroughs repeatedly. It provides the foundation for a company's ongoing renewal and growth. A firm with this capability has more operating than reliance on hero scientists, strong champions, or mavericks for breakthroughs every once in awhile. There is a system built into the company's infrastructure that addresses uncertainty and risk. It is different from the mainstream management system that's focused on the relatively low-uncertainty world of operational excellence and maintaining customer loyalty. It is, in essence, a management system for innovation.

What We Mean by a Management System

To be successful, any established organization must have a set of systems, structures, and processes that allows it to function efficiently and effectively. Typically when companies are founded, they are run by the founding entrepreneur or the founding team who makes all decisions, writes every check, draws up invoices, hires all the people, and assesses their performance. Obviously this works for only a very short period of time. An organization that does not adopt mechanisms for managing routine activities finds its opportunities for growth hampered.

A *management system* is that set of elements needed to make an organization function effectively and efficiently. It moves decision making and execution beyond the original founder and ensures that behaviors are oriented to achieving organizational objectives. The five necessary elements are shown in Figure 1.1.

Figure 1.1 Management System Elements

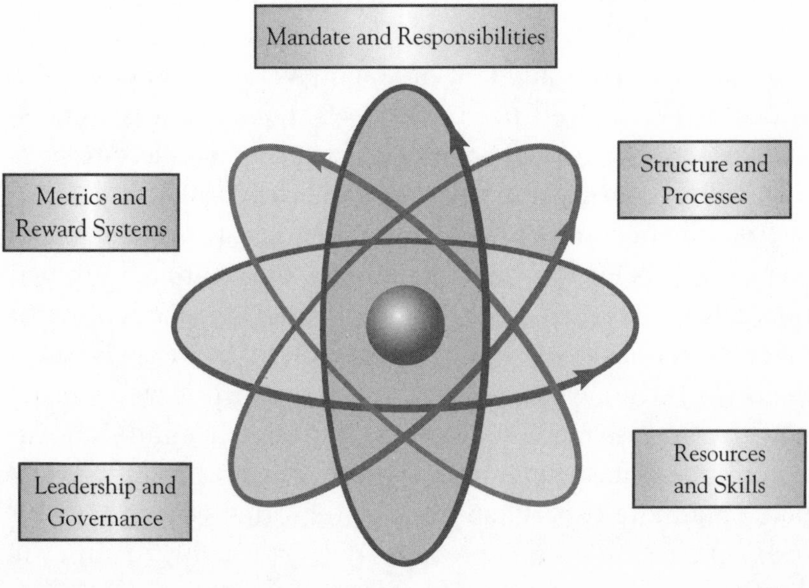

Mandate and Responsibilities

Structure and Processes

Metrics and Reward Systems

Resources and Skills

Leadership and Governance

Mandate and Responsibilities. The system's objective or mandate, the first element, is about its purpose and what it is responsible for. The management system for ongoing operations of any established company must efficiently and effectively manage current markets and operations, so as to responsibly leverage stakeholders' investments into profits. An innovation management system has a different purpose: creating new businesses as platforms for the firm's growth and future health.

Structure and Processes. The second element is the organizational structure and processes designed for innovation. Is there a group, department, or division responsible for breakthrough innovation? To whom do they report? Where is the locus of innovation activity? Should the innovation system be organized hierarchically, or should it be flat? Is it centralized or decentralized? Formal or informal? Rigid or flexible?

What about innovation-related processes? In an ongoing operation designed to efficiently and effectively deliver goods and services in response to customers' needs, processes abound. Market research, production scheduling, purchasing, inventory management, supply chain management, capital equipment maintenance, and new product development and launch all are driven by well-hewn processes that have been refined with experience. Indeed, ISO 9000 and other manifestations of the quality movement in the 1980s and early 1990s are a testament to the importance of processes in terms of cost savings and quality improvement. But processes for breakthrough innovation defy the concept of step-by-step variance reduction that is inherent in the process improvement techniques we apply today. Every breakthrough innovation requires learning many, many new things. The familiar can, in fact, be detrimental. Innovation processes must differ from those for ongoing operations and take on a more experimental, learning-oriented nature. This does not mean to say that processes don't exist, only that they're different.

Resources and Skills. System resources fuel the management system. How does the system access the resources it needs? Ongoing operations are self-funded, generally and are expected to generate returns beyond their expenses. Innovation systems, however, are investments and must receive resources from the larger corporation. Are the system resources provided consistently or in an ad hoc manner? Are they contingent or stable? Are they considered an investment or an expense?

What about finding and developing the necessary skills for innovation? The types of skills and talent needed to accomplish the system's objectives, and the definition of roles and responsibilities, as well as the mechanisms for developing and promoting that talent, must all be in place in order for any management system to function effectively. The skills necessary for innovation differ from those required to run an ongoing operation, and so one would expect to find different roles, career paths, and recruitment and development strategies for people in an innovation system than for those rising through the ranks of the mainstream organization.

Leadership and Governance. How are decisions made? Who makes them? Leadership for ongoing operations is oriented toward execution of current plans, working to prevent any deviations. Decisions are made by those in positions of authority based on predefined sets of criteria. But leadership for new business creation and breakthrough innovation must set a culture that tolerates learning and experimentation, creativity, failure, and the parsing together of tidbits of information to chart a new course on a regular basis. Decisions may be made by a broader set of people, since gaining buy-in for these new businesses may be a strategic choice that involves a number of constituents in the company.

Metrics and Reward Systems. Metrics used to measure the system's performance and the reward systems for those operating within the system are the final element that rounds out the management system. In an ongoing operations system, individuals are rewarded for executing according to a plan, without deviating from the budget or the schedule. They are paid bonuses for following directions.

In an innovation system, where the likelihood of failure is high, it's inappropriate to reward for sticking to a plan since plans are upended regularly. What is the best way to reward those willing to take the path of uncertainty? Should breakthrough innovation teams have equity stakes in the ventures? Should the breakthrough innovation management system be measured on how much money it brings to the company? Over what time period? Adopting traditional near-term profit objectives for an innovation management system clearly would be foolish.

How a Successful Management System Works

A management system cannot be successful unless its elements reinforce one another. For example, if the decision-making criteria used to evaluate projects for funding are based on what is already known about success in familiar markets and with known technologies, but the projects being evaluated are characterized by high uncertainty and ambiguous outcomes (Will the technology work? What are the most likely applications? How might we derive value from this as a business? How will we develop the process innovations necessary to make this economically justifiable?), it's very unlikely they'll be funded. If the system's objectives are to commercialize breakthrough opportunities, the decision criteria used must align with those objectives. Similarly, if the system's objectives are operational excellence, hiring people who are highly creative but who struggle with sticking to a decision would be a disaster. Yet those very same people may thrive in an innovation system, where exploration and experimentation are highly valued.

The management system for mainstream operations must differ from that of the breakthrough innovation function. Although the components of the system are the same, the way they operate is quite different, as illustrated in Table 1.2. The process for new product development in current lines of business fits with an

Table 1.2 A Comparison of Mainstream and Innovation Management Systems

	Mainstream Management Systems	*Innovation Management Systems*
Objectives and mandate	Efficient, effective management of current markets and operations	New business creation in new and existing markets
Leadership and culture	Planning and delivery oriented	Learning and building oriented
Structures	Clear and delineated	Flexible
Processes	Stage-gate, project management oriented; avoid deviations from budget or schedule	Learning and experimentation oriented, allow redirection based on new insights
Governance and decision making	Go-or-kill criteria clear in advance, hierarchical decision making	Decisions made based on strategic intent and continued learning; criteria not clear in advance; governance rather than hierarchy
Skills and talent development	Functional expertise	Entrepreneurial expertise
System resources	Annual budget allocation	Resources acquired through many avenues
Metrics	On-time delivery, cost containment, profitability	Portfolio health and balance; connection with strategic intent of firm; new domains accessed; new resources garnered; new business starts

operational excellence management system. It efficiently leverages what the organization knows for responding quickly and effectively to customer needs or competitive threats to current product lines or markets. But for breakthrough innovation, the company must search for and create new knowledge, as well as leverage what it knows into new domains and develop new competencies to fill gaps as it goes. That's a learning-oriented, experiment-based exploratory system, which is not traditionally rewarded in cultures of operational excellence.

The point is that most successful firms have become adept in fine-tuning their mainstream management systems. But they're telling us that they need new avenues for growth. Top-level managers are turning their attention to developing innovation systems that can be sustained, so they don't have to rely on the one-off breakthroughs that occur by happenstance and strong personalities alone. Our research program has followed these efforts, and we've learned not only about what firms are doing but how they can do it better.

Defining a Breakthrough Innovation Capability

A breakthrough innovation capability comprises three distinct building blocks, their interfaces with one another, and their interface with the rest of the organization. All of this is depicted in Figure 1.2. Let's start with the building blocks: discovery, incubation, and acceleration. These are shown in Figure 1.3.

How convenient that we can use the acronym DNA for discovery, incubation, and acceleration. (We know; we're cheating a bit on iNcubation.) DNA, the biological sort, contains the genetic instructions for the development and functioning of organisms. DNA is the blueprint for an organism and its behavior. Similarly discovery, incubation, and acceleration are the building blocks of the innovation function in companies, Together they comprise an adaptable model for innovation in an organization.

Discovery is the creation, recognition, articulation, and elaboration of opportunities. It encompasses many activities, including external scouting of technologies and hunting within the organization for good ideas and scientific research, all to find and generate ideas. But there's more. Ideas are great, but they have to be

Figure 1.2 Model of Breakthrough Innovation Capability

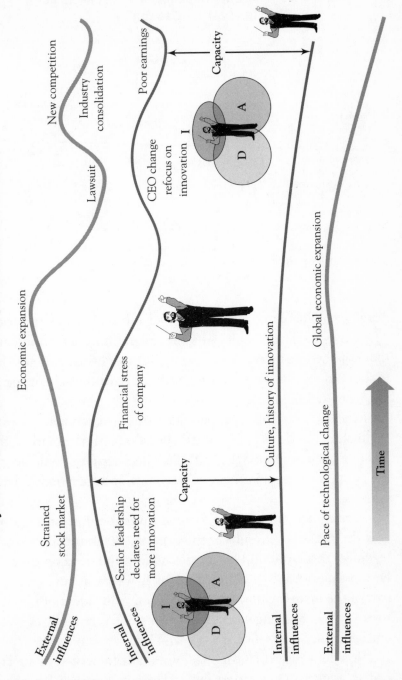

Figure 1.3 The Building Blocks of Breakthrough Innovation Capability

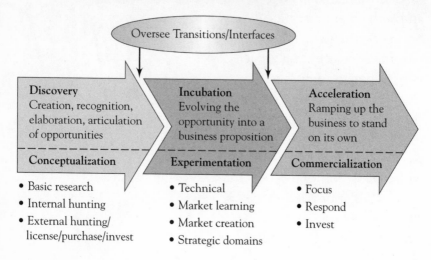

developed, elaborated, and envisioned as business opportunities. These may be flickers of insight or combinations of technologies that together create enormous possibilities for bringing new value to the marketplace. Discovery requires creativity and a high degree of conceptual skill.

Incubation is all about experimentation—experimenting with the technical side of opportunities but also with the market, the possible economic models that a fledgling business could adopt, the business strategy, partners, value chains, and operations-related options.

Acceleration refers to the focused investment on stimulating growth. It takes the results of the experiments in incubation and leverages those to build businesses rapidly and develop the new business platform to a point of maturity where it can survive as part of the mainstream company. If a company does not have all three of these building blocks in place, its BIC is not fully evolved, and something will suffer.

Now let's return to the idea of a management system. The management system for innovation plays out differently for each of the building blocks, as shown in Figure 1.4. The skills and metrics for discovery, for example, are different from those of incubation, and both differ from the skills and metrics needed for acceleration. So too do their objectives differ, their funding models, and their governance mechanisms. Each has its own expression of the innovation management system, but together they combine to form the company's innovation function. Also notice that Figure 1.4 shows that DNA is not simply a linear system through which a new breakthrough business opportunity progresses. We think of discovery, incubation, and acceleration as sets of activities, each with its own portfolio of opportunities ongoing within it. Resource allocation, prioritization, pacing, and other portfolio-level decisions will have to be made within and across each building block. This means that D, N, and A can't be managed separately; rather,

Figure 1.4 Management Systems for Discovery, Incubation, and Acceleration

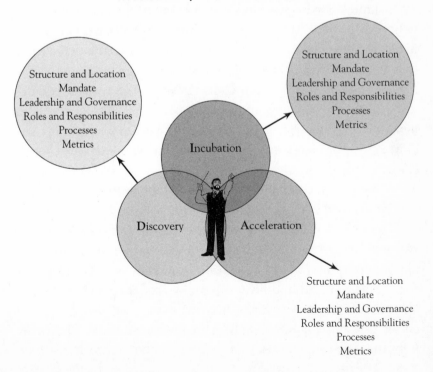

attention has to be paid to their interfaces. How will projects be moved from discovery to incubation? From incubation to acceleration? From incubation back to discovery if necessary? And how will new opportunities that become apparent in acceleration be captured in discovery? Someone must have responsibility for the overall function.

Finally, the model in Figure 1.2 shows that the DNA system occurs in the broader context of the company, the industry, and the economy. The DNA's interface with the rest of the organization will change based on these pressures that the organization is facing. We call that the organization's *capacity* for innovation, the subject of the next chapter. As Figure 1.2 shows, capacity changes over time.

Note the person in Figure 1.2 who is running interference between the innovation management system and the mainstream organization. The interface must be orchestrated (that's why our figure shows a person with a baton) by either a single person such as a chief innovation officer or a board. Someone has to be held responsible for innovation in established companies. "If someone doesn't own it, nobody owns it," the saying goes. History shows that new business incubators, new ventures divisions, or other groups that have been founded to accomplish major innovation in companies often don't last. Nortel Networks' didn't. Lucent's didn't. Xerox's didn't. Even 3M's didn't. Their average life span has been four years in fact.[9] The idea that an orchestrator powers the innovation function up and down given the company's ability to absorb new businesses is key to ensuring that the innovation capability is not lost, but remains part and parcel of the company, just as the marketing function remains even when marketing budgets must at times be cut.

The rest of this book describes each of the pieces of this innovation function in detail. We provide examples of how firms are approaching discovery, incubation, and acceleration, knitting them together and handling the tricky aspect of orchestration. Not one of our participating companies is truly satisfied with its current systems, but many of them are making progress. They're in it for the long run.

2

ASSESSING YOUR ORGANIZATION'S CAPACITY FOR BREAKTHROUGH INNOVATION

Rita (a fictitious name but a real person) was an assistant manager of project engineering for a large multidivisional firm's corporate engineering function. While attending an annual professional meeting of R&D managers, she heard a speaker on the subject of managing for breakthrough innovation (BI). Rita became aware, through further reading and correspondence with the speaker, of the challenges and different management approach required for managing for breakthrough innovations compared to traditional project management. She was enthralled with this new approach, which she believed could help her company plow new ground and move away from the faltering, risk-averse approach it traditionally took toward innovation. She described what she'd learned to Dave, her immediate boss, and he too became excited. Together they sought a sponsor in the senior management ranks to help them start an innovation hub. Through this sponsor (who was technical assistant to the group vice president of one of the company's largest groups), they requested the approval of the company's president. The word from him was "go do it." In fact, the president agreed to have regular meetings with the small team.

They received some initial funds, and off they went, assembling the core staff team they needed, facilitating idea-generation workshops, and developing a pipeline of innovation projects that didn't fit neatly within the company's lines of business and could dramatically change the industry in which they were participating.

Initially things went very well. They were finding people in the organization who were enthusiastically volunteering to work in their group. Employees in the business units were initially shy about stating their ideas for potential game-changing businesses they believed the company should initiate, but these idea-generation workshops really gained steam. People became enthused, and the group's reputation spread. People from central R&D began to seek them out for business development help. In fact, they were becoming overloaded with work. The opportunities were great.

Before too long, several of their projects matured to the point that larger investment amounts were needed. The group also needed a dependable, multiyear budget, since their portfolio of projects spanned longer time frames than the company's traditional planning cycle. They also needed to add more people (they were a group of about five at the time). Some of the projects were evolving in a manner that did not fit into the company's current business unit structure and would require some major organizational decisions.

Then the company merged, and a lot of change was underway as the two companies worked to become one. The president became increasingly unavailable, and controls grew tighter. The group was given less and less discretionary money, and pressures to get one of its businesses launched grew.

Rumors began to float that the president was going to bring in a new leader for innovation. This, in fact, happened in the group's third year of operation, when he named a vice president of new ventures and BI. This person had had significant experience in the venture capital industry, starting companies from scratch. But he had not had experience as a member of a large, established company.

Rita and Dave's group was shut down within six months; only one of their projects remained. The new vice president brought in three projects that he'd known of from previous consulting experience with the firm and formed a ventures division. Those projects remain in that division today. There is no pipeline. There are no staff members cultivating new opportunities, interacting with R&D or externally, or incubating potential businesses. Why not? Because the organization was not ready or able or willing to handle BI at the time. It had a constrained capacity for BI and, in fact, the more the president saw and learned of it, the more uncomfortable he became. What he wanted, in fact, was better incremental innovation.

Between the president's hands-off approach, his liaison to the venture capitalist consultant, the merger, and the company's culture of winning through dominance of current markets, the context was not munificent enough for the innovation hub team to get their legs. They'd gotten a great start, but did not have the capabilities to convince the senior leadership team that innovation from within was vital to the company. They clearly heightened the president's awareness of the benefits of innovation, and today they have four ventures growing in a separate organization. In addition, Rita now leads an innovation task force that is facilitating the business units' innovation objectives and developing a strategy for innovation using partnerships, cross-linking opportunities across business units, and improved incremental innovation. Who knows? Someday this company may decide it has leveraged incremental innovation to its fullest and will look for the breakthroughs. If Rita and Dave are still there, they may be able to restart their innovation hub. But perhaps not.

Companies cannot, and perhaps do not need to invest in BI at the same levels all the time. It's certainly true that companies vary their investments in marketing or R&D as required by competitive conditions or current strategy. But companies always invest something in marketing or R&D. They never expunge it from their organization; they just manage it differently. So too with innovation. We call this the organization's *capacity* for innovation. Depending on your company's capacity at the time, the innovation function must be managed differently. But it should never be extinguished. That's just too costly.

Capacity for Innovation

We define *organizational capacity* as the context and conditions for BI in the company at a given time. This includes the ability and will to resource and pace breakthrough innovations. Obviously capacity will vary. It is influenced by both internal and external factors. Some factors that influence capacity are fluid, meaning they change frequently, and others do not. Take a look at the model again, reprinted here as Figure 2.1.

Notice the two sets of lines that form a boundary around the DNA system? Those represent the company's capacity. The outer lines, on both the top and the bottom, represent the factors *external* to the company that are important to its ability and will to engage in BI, such as economic cycles, the entry of new competitors, or the pace of technological change. These factors cannot be managed, only anticipated (most of the time) and responded to appropriately. Examples of external influencers on capacity include:

- A multi-billion-dollar lawsuit levied against one of our companies in the 1990s prevented attention to the matter of innovation until it was settled in 2002.

- When the telecommunication market imploded in 2000 and, with it, the demand for fiber-optic cable, Corning's revenues plunged by 50 percent in one year, from $6 billion to $3 billion. Its stock dove from a high of $113.33 on September 1, 2000, to a low of $1.10 on October 8, 2002. It dampened Corning's ability to invest in innovation.

- The aftermath of the September 11, 2001, terrorist bombings threw the U.S. economy into a downward spin for approximately one and a half years.

- The methods used to sequence the human genome have provided opportunities for directed drug discovery and new understanding of diseases.

Figure 2.1 Model of Breakthrough Innovation Capability

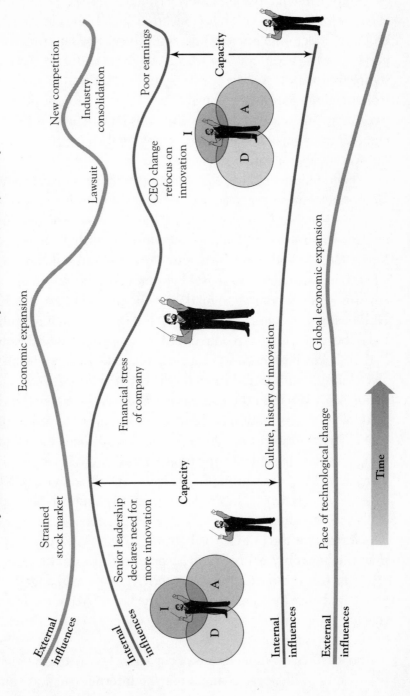

External influences

Strained stock market

Economic expansion

New competition

Industry consolidation

Internal influences

Senior leadership declares need for more innovation

Financial stress of company

Lawsuit

CEO change refocus on innovation

Poor earnings

Capacity

Capacity

Internal influences

Culture, history of innovation

External influences

Pace of technological change

Global economic expansion

Time

All of these factors are external influencers on a company's capacity for innovation. They come from outside the organization and may affect only one organization, the entire set of industry players, or the economy as a whole. They may affect the capacity for innovation positively or negatively. But they're not controllable and cannot be stopped. They have to be considered as a context. The most a company can do is remain aware of them, consider their impact on innovation, and set their expectations for innovation appropriately given the context.

The inner lines in Figure 2.1 represent *internal* factors. Internal factors are company specific and are brought on the company by itself in some way. For example, Jerry Junkins, president and CEO of Texas Instruments (TI) died suddenly on a trip to Europe in May 1996. He had been the driving force behind innovation at TI and, in particular, the digital light processing (DLP) technology that was just coming to market at the time of his death. The DLP business accounts for nearly $2 billion in annual revenues today, but at the time it required Junkins's support and leadership. Perhaps a key business unit's earnings plummet due to a recall or a key business unit introduces a product that's raking in very high margins. Or you're in the commodity chemicals business, so the focus of your organization is always on cost reduction rather than other sorts of innovation. Or you've historically been a company that focuses on big engineering projects, and you've been successful at that. All of these examples are internally based influences on a company's capacity. They're not necessarily controllable, especially over the short term.

Notice that the lines at the bottom of Figure 2.1 are less wavy than the lines on the top. That's because the lines on the top represent factors (both external and internal) that can change easily and may change frequently. Those on the bottom represent factors that are comparatively stable; they may change in ten- or twenty- or even fifty-year time horizons.

So we have two dimensions of capacity. The first is the *source* of capacity influencers. Sources may be internal (such as a lack

of appropriately skilled people or a CEO who is extremely supportive of innovation) or external (such as the general economic
conditions, or a new strategy from a competitor). The second
dimension to consider is the *dynamism* of capacity influencers.
Some may be highly dynamic and change very quickly, such as
competitive prices or a commitment to innovation as a result of
a senior leadership change. Other capacity influencers may take
many years to evolve, including the degree of industry turbulence.
Turbulence in the elevator industry, for example, is very, very low
and has been for many years.

If we consider these two dimensions together, an interesting
framework emerges, shown in Table 2.1.

Externally Based, Highly Dynamic Influencers

These are easily recognized, and there's not much anyone can
do about them. Stock market volatility plagues innovation, and

Table 2.1 Sources of Capacity Influencers

	Internal	*External*
High Dynamic Capacity Influences	Leadership changes, merger, number of breakthrough projects in late stages, stated business innovation mandate and scope, status of core business	Stock market, venture capital environment, lawsuits, changes in regulations, new competitors, technology landscape, currency exchange rates, availability of appropriately trained workforce to draw on
Low Dynamic Capacity Influences	Nature of technological innovation historically pursued, commitment to a particular product architecture, degree of capital intensity, belief system about innovation, organizational culture and values	Degree of industry turbulence, strength of external networks, market structure, regulatory strength, trade barriers, presence of industry consortia conducting research or promoting innovation-friendly government policy

venture capital availability for new business creation outside the firm can create more start-ups to invest in. Regulatory changes, new competitors, and novel technical discoveries can all sway the firm's ability and will to persist in developing its BI portfolio.

Externally Based, Low Dynamism Influencers

These influences can't be changed readily. The elevator industry hasn't had a major innovation in over a hundred years, the people at Otis Elevator told us when they developed the solution to the problem of limits on elevators in mile-high buildings. Some industries, such as the nuclear power and electric power industries, are run as oligopolies for years. Others, such as the defense industry, are plagued by regulatory oversight. Still others, like the pharmaceutical industry, compete on the basis of technological innovation and discovery.

Internally Based, Highly Dynamic Influencers

These influences on a company's innovation capacity include senior leadership turnover, specifically cases in which the new leader places a new priority level on innovation. Examples include GE's Jeff Immelt, who succeeded Jack Welch in 2001. Welch had started out as an entrepreneurial sort, but he migrated quickly to an operational focus on process excellence and growth through acquisition. While acquisitions continue to play a role in GE's growth, Immelt recognized the need to return to innovation for GE's next waves of growth, and the company has completely refocused. Similarly, the return of Corning's CEO, Jamie Houghton, from retirement to replace John Loose in April 2002 refocused the company on innovation. Mergers, like the Sealed Air and Cryovac merger in 1998 and GE's acquisition of Amersham influence a company's innovation capacity. Cryovac, it turns out, had a history of organic growth through technological innovation, whereas Sealed Air had grown primarily by acquisition. Through this partnership, Sealed Air gained increased capacity for innovation.

The mandate for BI in a company can change on a whim as well. First, it's supposed to focus on opportunities that are far into the future but aligned with the company's current business units. Then it's supposed to focus on white space opportunities: those that are adjacent to current business domains. Then it's supposed to be focused on multialigned, or system-level, opportunities. And then it's supposed to be project focused, then platform focused. It's enough to drive anyone crazy.

A final example of a highly dynamic internal influence is the status of the core businesses. The financial health of the operating companies has a clear impact on its ability and will to provide resources to and absorb new innovative businesses. It's not always straightforward how, though. Companies that are swimming in money may not feel the sense of urgency needed to pay attention to growth opportunities. One of our company business innovation portfolio managers told us, in fact, that his greatest anxiety was that the senior leadership team would grow bored with the projects that he was incubating. Business was good, and top management needed to ensure that it would stay that way by making investments to ensure the growth and health of domains in which they already played. Here the excitement was in the growth of existing businesses rather than in investing in new businesses with returns far in the horizon.

Internally Based, Low Dynamism Influencers

These influencers of capacity are listed in the lower left cell of Table 2.1. The first is the nature of technological innovation that historically has been pursued. If the company has traditionally competed on process cost control or if it considers itself a materials company but has no expertise in or willingness to migrate to other parts of the value chain, for example, its innovation capacity may be constrained. Another internally based capacity influencer that cannot be readily changed is the company's reliance on fixed assets, particularly those that are highly specific in terms of their functions. Albany International, for example, makes the fabrics that convey the paper sheet through a paper-making machine when converting wet pulp into paper. Those fabrics are woven, like big blankets. The looms required to weave the fabrics are useful only for that purpose and are extremely expensive. For Albany International to dispense with that equipment in order to invest in a prototype development lab, for example, would be impossible. Similarly, Boeing's investment in physical assets is enormous and specific to its purpose. Services companies may have greater freedom in this regard. The investment in physical plant or assets is much less of a constraint on investment firms, for example, which continuously consider the next financial instrument that could create breakthrough value for the marketplace.

A company's belief system about innovation operates as a capacity influencer, and we note that this is such an embedded part of a company's culture that we list it as an influencer that does not change easily. Gary Einhaus, R&D director at Eastman Kodak and one of the cofounders of the Systems Concept Center there, clarified how a belief system that values innovation operates:

> It has to be something that is almost a religious belief. This is not instant gratification. That's one of the reasons why people get excited about cost cutting, right? Because you take an action and see the results immediately. Breakthrough innovation is not like that. You don't get instant gratification. It's got to be a belief system, and you have to stick with it. You've got to put systems in place that enable it, and you've got to believe it's going to be successful. And if it's not working, then go in and interrogate why it's not working. Don't stop it.

Finally, organizations have cultures that enrich or constrain innovation. Google, for example, is famous for encouraging a culture of creativity and invention through signals it sends to its employees. Dress code, organization structure, office design, meeting protocols, and the norms for addressing one's superiors are just a few of the cultural cues you might consider. These do not change rapidly in an organization, because employees are attracted to work for organizations whose cultures fit their own values and norms, so the culture may become increasingly entrenched, unless leadership is installed from outside the organization.

Summarizing Capacity

Although there's not a lot you can do about capacity influencers, it's important to be aware of them, to think about their source and their duration. We'll see in Chapter Seven, on orchestration, that assessing your company's capacity is key to determining the best way to initiate building a breakthrough innovation capability and certainly key to sustaining it over time. The approaches companies have used in the past (intrapreneuring, skunk works, corporate venture capital, internal corporate venturing) can all work but under different capacity conditions. A few years before CEO Jerry Junkins's untimely passing, one of us engaged in a long discussion with him concerning innovation at Texas Instruments. He indicated that many of the standard business practices of U.S. companies, and in particular at TI, had been and were a major detriment to ongoing innovation drives. "First," he said, "in slow-down periods we lay people off and cut back on our activities, including innovation efforts. When things pick up again, we try to pick up where we left off and ramp up our innovation efforts to help ensure a solid future. We try to hire back the best people that we laid off but usually it is too late; they have gone on to something else. We consequently lose a lot of knowledge and learning and cannot get it back. This has been all wrong. Slow-down periods are opportunities to improve, to acquire new capabilities, new opportunities, and to gear up for the future health of the company. This we observed from the Japanese when they were dominating the manufacturing landscape. Second, we are much too successful in our innovation projects. Our success rate is well over 90 percent." When asked to elaborate, Junkins expressed the belief, born of personal and company experience, that "we have such a high success rate because we are only pursuing easy and safe innovations. We are not taking enough risks, and we are not taking on enough uncertainties. We only become good at learning when we learn how to take on uncertainty on its own terms. Otherwise we become stagnant. Our future will be compromised if we are not in pursuit of major breakthroughs and the difficulties that they bring. We may not be so successful in our pursuit of major breakthroughs that we will be first and foremost when they eventually do occur, but we will be in the game and we will get a share."

Munificent and Constrained Capacity: How Do They Feel?

Next we provide two detailed examples of organizational capacity from the companies we studied. We want you to experience (vicariously, of course) a company with a rich capacity for innovation and one that currently has a constrained capacity. You'll realize that these can change, and indeed may already be changed between the time we write these words and the time you read them, but the importance of these examples is to show how different companies' capacities can be. After that, we'll discuss the implications for action.

General Electric Company: A Case of Rich Breakthrough Innovation Capacity

For nearly a hundred years, the General Electric Company led the way in developing some of the most advanced technologies and innovations that the world had ever seen.[1] Then from 1981 to 2001, under the leadership of Jack Welch, the corporation witnessed a dramatic shift in its culture. It went from a company focused on innovation to one that increasingly rewarded operational excellence. Mergers and acquisitions were the order of the day. Nonorganic growth was emphasized. Growth by acquisition became the mantra, and by all accounts, this strategy was highly successful.

In September 2001 when Jeffrey R. Immelt was appointed CEO, he decided to change the rules that characterized Welch's success by going back to the legacy of GE's founder, Thomas Alva Edison. The new focus for GE was innovation, and it became the mission for the new GE: long-term funding for research and development, organic growth, and the development of breakthrough technologies. According to Immelt, the central R&D group had been so focused in the Welch years on servicing the immediate and near-term needs of the business units that long-term research and organic growth through BI were no longer part of their fabric. He said:

> I started my career selling, and I made this profound discovery that whenever I had good products to sell, I always did better than when I had lousy products to sell. So we're closing the door on a decade that was about capital markets and acquiring things and opening the door on a new period that's more about developing things. The companies that know how to develop things are ultimately going to create the most shareholder value. It's as simple as that.[2]

It was not going to be as simple as that, however. GE was, and still is, a highly diversified company with 2006 revenues nearing $132 billion and businesses in power generation, medical imaging, aircraft engines, appliances, and financial services. Creating new technologies on such a large scale was going to require a large organizational infrastructure and competencies for innovation. In a global organization such as GE, the execution of such an endeavor would require funding and resources on a large scale that would be sustainable for the long run.

Major investments were made in corporate research and development. Approximately $100 million was spent on revamping the Global Technology Center in Niskayuna, New York. Three new technology centers were established across the globe. The John F. Welch Technology Center, the second largest in the world, was established at Bangalore, India, in 2000; the Technology Center in Shanghai, China, opened in 2003; and the Technology Center in Munich, Germany, began operations in 2004.

Each center was to be a highly specialized technology node and become a world leader in its domain of expertise. Each of these centers employs hundreds of scientists and engineers and has made large investments in laboratory and other research-driven infrastructural facilities. By September 2003, just two years after Immelt became CEO, $400 million in new money had been committed to the development of the global technology centers. The company had demonstrated strong financial performance for the five years since Immelt took over as CEO.

Another one of Immelt's initial signals regarding his seriousness about innovation as a key priority at GE was his placement of Scott Donnelly in the role of chief technology officer (CTO) and senior vice president of GE global research to head the global technologies centers. Donnelly's first move as CTO was to convince Immelt that an investment in a set of advanced technology programs was necessary to develop the leading-edge technological prowess that GE would draw on over the next several decades, consistent with Immelt's vision to create new platforms for the company's future growth.

The advanced technology programs (ATPs) at GE Corporate Research formed the core of the company's new initiatives for BI. Each program had its own group of highly skilled scientists and engineers and a program leader. Donnelly was responsible for the overall charter of the division's ATPs, which he initiated within the first several months as CTO, based on twenty proposals that R&D scientists wrote in response to a call for proposals that he issued throughout the GE research community. The following advanced technology programs were put in place at the research center:

- Nanotechnology
- Photonics

- Advanced propulsion
- Molecular imaging
- Biotechnology
- Light energy (organic light emitting diodes [OLEDs] and photovoltaics)

Each of these, Donnelly explained, was a completely logical play for GE. He described the ten-year future in terms of what these technologies would enable and why GE was strategically willing to invest to become a forerunner in each arena, based on its current capabilities, its vision of the future, and its drive to develop new competency arenas. He also increased the number of business unit representatives who were resident in R&D, to meet with the ATP teams and others to scout technologies that could be developed into nearer-term business opportunities.

According to Tom McElhenny, a business development manager within the central R&D organization, the ATPs were typically funded at $4 million to $10 million and received sustained funding for two to three years.[3] The programs were called blue sky projects since the results were not expected to be converted into commercial gains in the short term. Donnelly had this to say about the ATPs:

> Those are the kind of things that will be a big impact ultimately to the businesses. Those are the ones that we think are worth putting a bet on. And so that's what I say; we've got about 20 percent or so of the R&D that we spend in those kinds of programs. Because you can't do a reasonable market study. They're just not mature industries. They're certainly not mature technologies by any stretch of the imagination. But when I look at them, year on year, what I'm looking for is that kernel of something that we've discovered or think we can apply for, that we can take and spin into something where we would do a real market study.[4]

Donnelly met regularly with the senior leaders of each business unit and Immelt to communicate progress on the ATPs, identify new business possibilities, and plan the steps that each business unit should be taking, simultaneous to the technology development in the research organization, to develop the business and market receptivity that would be needed.

At the same time that the technology centers and increased funding for research were being implemented and the business unit leadership was preparing for the results of the ATPs, a number of other initiatives were also put into place, including these:

- *Imagination breakthroughs.* In 2005, Immelt had engineered a scalable and quantifiable process for coming up with money-making "eureka" moments.[5] His high-profile "commercial council" consisted of top divisional heads and senior sales and marketing executives. The council met every quarter to evaluate ideas from the senior ranks that aim to take GE to new heights of achievement. Some of GE's cross-functional structures for innovation are listed below. These were large projects and typically were required to take GE into a new line of business, geographical area, or customer base. At least three were mandated per year, each was required to generate $100 million in incremental growth within three years, and each was scheduled to receive billions of dollars in funding over the coming years.

- *Idea jams and a virtual idea box.* Executives from different divisions meet face-to-face or virtually to brainstorm and generate new ideas.

- *Ideation courses.* Designed to spark idea generation, these courses were available to any employee in the company.

- *Excellerator awards.* To encourage excellence in a chosen field across the organization and reward people for initiating great ideas.

It has become clear that Immelt has made technological innovation an integral part of GE's agenda. And in typical GE fashion, when a big initiative is committed to, there is a driving focus on excellent execution.

GE enjoys a rich business innovation capacity environment, which can be characterized as follows:

- Breakthrough innovation is visible, clear, and supported as an integral part of Immelt's agenda. He expects all members of the company to be involved in it, either as their primary job function (the ATP teams) or as a back-of-the-mind, ongoing responsibility.
- The CEO and CTO work well together. They share the objective of innovation and implement it in their respective ways, all the while supporting one another.
- There is a rich senior leadership network that supports the initiative on a daily basis. Immelt and Donnelly involve senior leaders in all parts of the organization, so that the strategic intent they imagined would have the commitment of others necessary to realize it.
- A great deal of money has been invested in the ATPs, along with the recognition that they may not deliver immediate returns.
- There is an excellent talent pool of young, brilliant scientists already in the R&D organization who have been willing to learn how to do business innovation more effectively.
- GE is not under financial duress.
- Innovation is reinforced through rewards and funding as a way of deploying a culture shift that supports it.

Table 2.2 applies the capacity influencers in Table 2.1 to the GE case.

It is clear that under this rich capacity environment, GE has made tremendous progress in implementing a set of activities and systems for enhancing BI. Not every company has the where-withal for matching the scope of GE's initiatives here, but in a rich capacity environment, there are opportunities within the scope of any business for increasing breakthrough innovation capabilities substantially. Now we turn to a different environment: a constrained

Table 2.2 Rich Capacity Influences at GE

	Internal Capacity Influences	*External Capacity Influences*
High Dynamic Capacity Influences	New CEO sees innovation as a basis for competitive advantage; current operations healthy; available talent pool has skills necessary for new business creation	New technological horizons emerging (for example, the human genome); global economic health; world awareness of the importance of alternate energy sources; market support for many of the Advance Technology Programs
Low Dynamic Capacity Influences	Financial stability, rich internal networks, history of technological innovation, flexibility of organizational structure, willingness to move into new application spaces, belief that innovation is the source of competitive advantage	Moderate degree of industry turbulence, U.S. national policy promoting innovation

capacity. While the case we describe is disguised, let us assure you that it is real and comes from one of our participating companies. We'll call them Diversified Industries.

Diversified Industries: A Case of Constrained Capacity

In contrast to Scott Donnelly at GE, George Madison felt frustrated.[6] As director of the Breakthrough Innovation Group at Diversified Industries, he'd been trying for several years to build momentum, a portfolio of rich breakthrough opportunities, and commitment from his leadership. The deck was stacked against him.

Diversified Industries, Incorporated is among the world's leading industrial companies, with 2006 revenues of more than $25 billion and nine thousand employees, focusing on the production of key commodity components and their delivery to large industrial customers. Diversified Industries had gone through a dramatic change since the mid-1990s in its R&D strategy and organizational structure. In 1995, nine of its U.S. labs were shuttered and its basic research budget was reduced by $50 million, essentially terminating most exploratory research activities. In 1998, Diversified Industries reorganized its business units, combining some and reducing others to go from a total of twenty-one to eight and then reduced once again to six. The purpose of this restructuring was to narrow the product portfolio and focus on basic commodity components and to align all innovation to the remaining business units. As a result, the sales volume of base components has increased substantially since 2000.

There was some concern, however, about Diversified Industries' new business creation activities during this period, because the company's primary emphasis lay on generating cash from existing markets with current technologies. There was concern about how to keep the research scientists motivated and how to coach inventors to articulate value propositions in the face of reduced R&D funding. This was the background behind setting up the Breakthrough Innovation Group in June 1999. The model for the group came from Diversified's parent company, which had supported a successful Breakthrough Innovation Group for many years.

George Madison, who joined Diversified Industries in the late 1980s as a research scientist after receiving a Ph.D. in physical chemistry, applied for and became group manager of this program. He built a team of five people, and they set about their mission to develop sustainable BI, all the while envisioning they'd be contributing to new profit resources and producing a potential "next big thing." To execute on this mission and vision, they initiated five programs, each directed by one of George's core team members:

- The purpose of the *internal ventures program* (IVP) was to provide a process and some resources for innovators to formulate, develop, and evaluate their ideas for innovations. The aim of projects in the IVP was to exploit known technologies for commercial gain in new markets or new applications.
- The *innovation lab workshop* (ILW) was regularly conducted in a lab setting to generate novel business or technical ideas that would be the source of new R&D projects. A structured approach to idea generation and development, its primary objectives were to generate and test proposals that potentially formed the basis of new profit sources and generate ideas for new experimental programs aimed at business renewal.
- The *advanced technology program's* purpose was to develop novel technologies that could create future opportunities for the company or develop responses to emerging competing technologies.
- The *external scan program* (EP) was one in which volunteers monitored key technology developments that were emerging outside the firm and considered ways those technologies could either disrupt the company's competitive advantage or provide opportunities for leaps forward.
- The *idea finder's group* shared ideas, trends, and contacts; identified discontinuities; and suggested new focus areas for proposal generation.

Each program had its own manager and overall responsibility was Madison's. Managers of each program were selected through the company's posting system, and all had technical experience and passion for this new activity. However, although these managers and Madison were experienced in managing resources in a business unit, they lacked experience in new business creation.

The Breakthrough Innovation Group had the capability to consider a wide range of potential opportunities. The CEO of Diversified Industries decided to have them pursue only opportunities that were aligned with current activities of the business units, since that was his goal for growth. In this context, the IVP, which acted as an internal 'venture capital' fund for ideas, used the following criteria for approving and funding projects: the opportunity must:

- Be well aligned with one of the existing business units
- Provide an opportunity for potential profitability,
- Demonstrate market share defense,
- Leverage existing or accessible competencies

If an IVP proposal was a strategic fit with business units and one of them deemed it interesting, it could be developed through the business unit's funds, with the goal of eventually moving from the Breakthrough Innovation Group to the unit that sponsored it. Although Madison and the IVP manager had the best of intentions for the venture fund and its small portfolio of companies, they had little support and no resources to help the fledgling opportunities. Neither one had built or run a business before. In fact, neither had ever worked outside the R&D setting. Nor did they have the discretion to hire anyone who did have that experience. So the fledgling ventures met with the IVP manager once in a while and reported progress to an oversight board, which was composed of Madison's team and some additional R&D directors. Still, no new business creation help was forthcoming.

Budgets for the Breakthrough Innovation Group's program were reduced from $5.5 million in 2001 to $4.2 million in 2002. With only six full-time managers and reduced funding, the opportunity for long-term innovation was clearly limited. The focus of the business units was near-term and midterm, making the most of the current operations. Few business unit leaders were thinking about breakthrough technologies. Even if a short-term project showed promise but it was not supported by the business units, it was terminated because there was no other place for the project to be supported financially or with expertise. The breakthroughs group could not propose the creation of a new business unit if they came on an opportunity that appeared great for the company but ill fitted its current organization structure.

Once the program was initiated, results did not meet expectations. People were not interested in working on the Breakthrough Innovation Group's projects, since the highly risky and new projects were not viewed as a path to promotion and not many projects were proposed. There was a greater reward for working on business unit projects rather than breakthrough projects. So the breakthrough group put into place a concept that they called "floorwalkers," whose responsibility would be to talk to scientists and engineers in the business units, understand their ideas, and help them submit their ideas in the form of proposals. Through this effort, the number of proposals to the IVP increased from thirty-nine in 1999 to fifty-four in 2001.

The breakthrough group worked passionately not only to develop new ideas and commercialize them but also to educate employees and create an innovative environment. They issued a periodic newsletter to interested employees, providing information on their portfolio of projects, novel technologies, and emerging trends. From 2000 to 2003 they invested in and developed twenty-six projects within the company and two projects with outside organizations (one university and one small company), and they initiated a cross-group project with other divisions of Diversified's parent organization, although there was no mandate from the senior leadership and no incentive for an individual to work on such an initiative.

The development of an innovation strategy started after the Breakthrough Innovation Group's program was founded. Paul James, vice president of technology, who was the group's most significant link to the Senior Executive Council, had been busy with this strategy rethink. James's criteria for new projects required a strategic fit to the company's overall operational excellence, process improvement model, which was in contrast to Madison's desire to pursue long-term goals and create the "next big thing" every decade. Instead of radical innovations, James thought the organization was more adapted to incremental innovations.

There were different views about Diversified's innovation strategy within the Breakthrough Innovation Group as well. Some of Madison's team were anxious about his belief that BI could happen in the company. They believed that the program needed to focus on current issues in the business units and solve those in order to gain, and maintain, credibility.

Then in 2003, the commodity component industry went into a decline, and resources at Diversified became increasingly scarce. To cope with this situation, Diversified reduced R&D expenditures and reorganized to consolidate management responsibilities and cost. The Breakthrough Innovation Group was combined with strategic planning to form the strategy and innovation group, which, as the name implies, combined a number of the activities that were related to the company's technology strategy and the interfaces with its overall business strategy. The Breakthrough Innovation Group was funded at $2.5 million out of the $100 million R&D budget.

CEO John Vanrose replaced Paul James as vice president of technology in November 2003. He used the strategy and innovation group as a think tank to develop scenarios of the future, in addition to maintaining the IVP internal venture capital fund. He did not, however, increase its budget or add staff to help coach the nascent business teams.

Of the Breakthrough Innovation Group's other programs, the externalization program was combined with similar programs in the German office and the hunter's network remained, but the others lost steam. The team of five began to drift apart. George Madison, after going through emotional ups and downs for four and a half years, moved on to a small division within the parent company, where there was a stronger commitment to growth. The manager of IVP decided to return to technical work. Two of the others found jobs with the corporate-level Breakthrough Innovation Group. One remains in the strategy and innovation group, focusing on long-term trend analysis. Table 2.3 summarizes some of the ways in which the capacity—in this case, constrained capacity—played out in the Diversified Industries environment.

What we see from this case of constrained capacity is an inability to gain traction for BI activities. Innovation is constrained due to a number of factors:

- The Breakthrough Innovation Group program was allowed because it was a function that existed at the parent company, corporate level, and so could not be actively resisted by Diversified's company executives, but it sure could be passively resisted or starved.
- The Breakthrough Innovation Group was used to prevent complete morale slump in R&D when the exploratory research budget was cut.
- The CEO wanted only aligned business unit applications.
- The CTO was oriented to process improvement in the service of business units. While there certainly can be, and are, breakthroughs in process innovation, they are rare.
- There was no budget to fund anything beyond discovery.
- The corporate culture was focused on cost cutting.
- There was low turnover, so the talent pool was stagnant.
- The company moved to a commodity orientation over the previous fifteen years and, in fact, had sold off higher-value-added businesses.
- The industry is heavily monitored.

It is no wonder that innovation did not happen at Diversified Industries.

Table 2.3 Constrained Capacity Influences
at Diversified Industries

	Internal Capacity Influencers	*External Capacity Influencers*
High Dynamic Capacity Influences	Downturn in core business; lack of appropriate knowledge, skills, and experience; senior leaders focused on today only	Economic recession, raw material stock shortages, lawsuit
Low Dynamic Capacity Influences	Low-cost producer; heavy investment in specific fixed assets; low turnover; low growth; theory of competitive advantage not based on technology or innovation; uncertainty, risk, and fear of failure among employees	Tightly controlled regulatory environment, entrenched industry standards, basis of industry competition is determined and well understood

There is a postscript to this story. The parent company has reorganized its entire company, including central R&D, to focus more heavily on white space innovation. Focus has returned to the Breakthrough Innovation Group, but at the corporate level. Capacity can change, so it's best to be ready.

The 'So What?' Question

All of this description must have you wondering what you should have learned from this chapter. We have three points.

First, when your company is experiencing a rich capacity, get busy building your discovery, incubation, and acceleration building blocks. It's hard enough to do this when times are good, but when times are tough, it's nearly impossible. George Madison tried and tried, but couldn't get the help he needed at the time. The danger is that you can fritter this time away by not taking the opportunity seriously enough. Use the opportunity to get the DNA set up, get the pipeline initiated, get the right people, and get one or two early successes (perhaps not commercial successes yet, but at least demonstrable progress). Your objective is to become mature enough as an innovation function and management system that when tough times come, innovation is established as a function in your company.

Second, when your company is experiencing a constrained capacity, it's all about small wins, aligned opportunities, co-opting key people, and exercising your networks to ensure you stay afloat. Measure what you've contributed to the organization so that everyone remains convinced of the value of the innovation function.

Third, there may be times when the organization has less than limited capacity for innovation. Perhaps there's a solvency threat or some similar life-threatening situation. We admonish you to keep the innovation system (people, budget, mandate) in place, but realize you'll have limited resources and attention.

Our point is this: rich capacity provides the opportunity to develop DNA competencies, but this can be squandered if there is no urgency to get DNA up and running quickly. Constrained capacity doesn't mean you can't make progress. It depends on the sources of the capacity and their dynamism. If it's something that can change within a reasonable period of time, you can begin to build constituencies, you can put your well-honed DNA capabilities on the back burner, and you can reduce the number of investments, but keep the system active. If the sources are unmalleable for long periods of time, the challenge is much more difficult.

Assess Your Company's Capacity for Innovation

Following are questions that must be asked and answered in order to assess the capacity of a firm to conduct BI. The answers may be immediately apparent, or they may require much detailed research.

These questions should be addressed relative to BI systems and relative to prospective BI projects, platforms, or portfolios. Addressing them can provide valuable insights as to whether to begin a BI project or platform; whether to start an initiative; and what approaches to take in building, experimenting with, monitoring, and developing your BI system. At the very least, capacity should be assessed in advance of every initiative and continuation decision to help you adapt your approaches to ever-changing environments and set appropriate expectations.

1. Is the company facing solvency or survival issues?

2. Is the company facing declining revenues or declining profitability issues?

3. To what extent does the company possess the enabling business, market, technical, and organizational expertise to pursue a BI over the course of an imagined business end to an actual commercialized business end?

4. What are the major drivers of growth in the company now and expected for the future?

5. To what extent are investments available for BI in relation to investments for existing businesses? Are the time frames sufficiently long for the innovation investments?

6. Breakthrough innovation is characterized by high levels of uncertainty, risk, ambiguity, need for learning, and need for experimentation. To what extent can your firm in general, the senior management in particular, and the prospective BI team tolerate these characteristics for the short and the long terms?

7. To what extent does innovation suffuse your company's culture?

8. To what extent does your BI team possess and have the ability to use rich, powerful, extensive, and extendable networks inside and outside the company?

9. To what extent can senior management be actively, visibly, and materially involved in the entire course of BI, from before initiation to full commercialization?

10. To what extent are learning and capability acquisition key parts of both strategy and strategic intent?

11. To what extent is your current strategy sufficiently invested in growth or renewal to support BI?

12. To what extent is the company consistent in its support for innovation?

13. To what extent does the company have the ability and the will to persevere under all the risks and uncertainties of bringing its prospective innovations to ultimate commercialization considering the potential hazards posed by business cycles, political actions, regulatory thrusts, competition, and industry environments?

14. To what extent are the prospective investment risks and rewards expected from your BI program commensurate?

15. To what extent can the BI domain in which you wish to focus be funded by external sources?

16. To what extent are there existing or impending regulatory barriers to the innovation domain where you wish to focus?

17. To what extent are there already good-enough solutions embedded in the marketplace for the innovation domain where you wish to focus?

18. To what extent does your innovation program seek to solve problems that are poorly addressed in the marketplace and for which the prospective market capture is sufficiently large to justify the investment and risk?

3

THE DISCOVERY COMPETENCY

The largest portion of industrial R&D budgets is spent on existing product lines or newly identified platforms; the proportion of expenditures allocated to fundamental research has been low. Only in 2004 did industrial companies shift the focus of their R&D spending from directed research and support of existing businesses to new business creation projects.[1] Savings in R&D expenditures are reportedly expected to come from reductions in the technical workforce, movement of technical development work to Asian countries, and leveraging resources through alliances and joint ventures with federal and university labs as well as smaller companies.[2] These trends signal that central research labs in established companies are less focused on discovering new laws of nature (that is, fundamental science) and are more focused on developing opportunities for breakthrough businesses.[3] In fact, that is what we've seen among the companies we've studied as well. R&D just isn't what it used to be. It's even more exciting now.

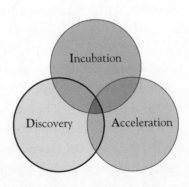

The first building block in our innovation system is discovery. A discovery competency is a company's ability to create and identify opportunities that may have major impact in the marketplace through the delivery of new performance benefits, greatly improved performance, or new ways of doing business. Breakthrough opportunities can come from technology push or a clear market need. They can originate from within the company or outside. The open innovation model,[4] in which new ideas, technologies, and opportunities are sought from sources and partners outside the company, is being practiced in every one of the companies we've studied. Discovery as we define it here is not equivalent to invention, it's not equivalent to creativity, and it's not equivalent to R&D. The reason it's not equal to invention is that companies can learn about phenomena in the world around them—inventions that may have occurred outside their organizations—and adapt them for their own purposes, in other capacities, or other applications, to boost their own businesses. Gregg Zank, the R&D director for Dow Corning's New Ventures business unit, explained that his group was focused on finding discoveries others have made and adapting them for Dow Corning's purposes. There was no need, he noted, for his R&D group to be conducting fundamental research itself. They had to know enough to recognize the importance and application of that research to the company's purposes. So the not-invented-here syndrome appears to be breaking down in most organizations today.

Discovery is also not equivalent to creativity. We say this for two reasons. First, creativity is required throughout discovery, incubation, and acceleration. Indeed, creative problem solving is one of the most highly valued skills in business today. But, second, creativity is traditionally thought of as that flash of insight, that Aha! moment, when disparate pieces of information are connected in someone's mind to provide a brand-new idea. Certainly that's important, but it's only a small part of the breakthrough innovation process or even of the discovery function itself.

Similarly, the reason discovery is not equivalent to R&D itself is that it's a broader activity encompassing more than science and technology exploration. Discovery incorporates a consideration of how an invention or an idea can be formulated into a business. It's all happening together in a number of the companies we've observed. Here's an illustration from one of them.

The inventor of a new technology, Paul, described his early experiences in the lab. He knew that his invention was exciting, but he had reached a stage in his technical exploration that he didn't know what to do next. He told us:

> I could've done what most research scientists do in that situation . . . ask for more money and a bigger lab. However, I realized I still wouldn't know what to do in that lab on Monday morning. I got forced out of the research lab because I didn't know who my customer was [so which business unit would take this on]. . . . The market didn't exist . . . so I couldn't develop the technology without studying the market. But if I wrote papers about the market, that wasn't really a technical research problem. So I wandered down the hall and described my discovery to Linda [who worked in the group responsible for nurturing breakthrough innovations in the company]. She asked me a lot of questions but convinced me that we had to start talking to other players in the market. We found and joined a university center that had sixty member companies whose purpose was to discuss standards in the intelligent packaging domain.

> I could've made up some stories and gotten a lab, a team, and a bigger budget. But it would be only because I didn't know what to do. The competitive intelligence we've gathered has convinced me that the places we thought this would work would not have worked out.

The senior-level champion of the project remarked to us how pleased he was that Paul had sought out Linda for help in formulating the opportunity that his invention formed the basis for: "This could have remained a research project, but it wouldn't have worked because it would've been treated as a technical problem. Really, though, it's a business problem."

A full discovery capability comprises these three sets of activities: activities related to seeking out and developing foundational knowledge in the basic disciplines associated with the domain, opportunity generation, and opportunity articulation. We will first look at two examples of rich discovery competencies in our companies, so you can see what a discovery competency looks like and how it plays out differently across companies. Then we'll get into the details of elaborating these three sets of activities.

Rich Discovery Competencies: Two Examples

A rich discovery competency is one that teems with urgency, excitement, and constant questioning of the company's potential role in emerging industries. It's not just about seeking opportunities but also about creating them—continuously.

Discovery at IBM

IBM has one of the biggest, best funded, most prolific research labs of any company on the planet. Indeed, year after year, IBM earns more patents than any other company.[5] Yet when CEO Lou Gerstner grew alarmed in September 1999 that IBM had missed the departure of the computational biotechnology train from the station, he did not call the chief technical officer. Instead, he shot an e-mail to three senior vice presidents: of strategy, of marketing, and of sales and distribution. His note asked, "Why do we keep missing these things???" After a probe lasting several months to find the answer to his question, IBM launched the emerging business opportunities (EBO) program. EBOs, also known as horizon 3 businesses, are "folios of experiments for long term growth."[6] These are new businesses that may not fit into the current organizational structure but nevertheless may be relevant to multiple divisions in the company. They require senior leadership investment and attention to ensure they'll be nurtured.

Identifying the first six EBOs was easy. The senior vice president of strategy and senior vice president of technology could list them in one short meeting. Each was a project or initiative that had been floating around in the company for awhile, but not getting traction because they didn't have the attention and resources they needed. Most were orphaned projects; they required resources from more than one business unit, and there was no mechanism to ensure coordination or shared funding from multiple business units in the company. Others required attention from units that were preoccupied with more immediate concerns.

Once they were identified, the senior vice president of strategy, the senior vice president of technology, and the assistant controller met with each young EBO team to help coach and advise them. In fact, that team of three very senior people met with each of the EBO teams for at least two hours each month, a total of sixty man-hours per month of senior leadership time devoted to long-range, breakthrough businesses. In addition to coaching the teams, the senior leaders made sure that there were resources to support them. They saw to it that money set aside in the divisions to develop the new opportunities was in fact invested in doing so and not diverted to solve the near-term crises that inevitably would arise. As a result, the EBOs flourished.

By 2001, the pipeline needed refreshing. Mike Giersch, vice president of strategy who oversaw the EBO processes, hired Debbie DeLoso to help identify new EBOs. She launched an all-out attack on generating new opportunities by engaging the company in a range of activities. She met regularly with R&D groups but also with others. She organized the "next big thing" conference, an internal meeting for members of the technical and business communities, as well as some outside experts including the venture capital community, which generated 150 ideas that formed the next set of corporate-level H3 projects. DeLoso developed networks within R&D as well as with division-level EBO leaders to decipher which new business opportunities arising in the divisions had cross-divisional application and should be elevated to corporate status.

By 2003, the EBO staff in the corporate strategy group was holding a different event every year for the purpose of identifying potential new horizon 3 opportunities, which they called idea jams, idea cafés, externally based trend analyses, deep dives, and so on, all to stimulate ideas for the EBO pipeline as maturing H3 businesses were transitioned on to H2 status. The intent was to bring a broad spectrum of people together in facilitated meetings for the purpose of surfacing new opportunities. EBO staff have tried a multitude of approaches, none twice so far. In fact, the freshness of the approaches each year helps to stimulate the big opportunities just as much as the fact that they engage in these activities at all. In other words, they haven't fallen into an idea-generation process trap of using the same techniques and the same people to come up with the breakthroughs, an approach that can get stale. Different approaches, different people, and the strategy and technology corporate executive team's eye on the outcomes all help the EBOs influence the company's articulation of where it wants to evolve to in the future—and so where to place its bets today.

The EBO program has been a rousing success. Twenty-two corporate-level EBOs have been through or are in the system. Twenty-five EBO businesses were launched in the first four years of the system's operation. Five of them together achieved more than $1 billion in revenue in 2003–2004. Linux, pervasive computing, and IBM's life sciences business were some of those initial EBOs. And the pipeline is continuing to be fed.

Corning's Discovery Competency

Corning Inc. boasts a rich hundred-year history of breakthrough innovation,[7] fueled by a deep technical bench (a total technical community of more than seventeen hundred employees, nearly six hundred with advanced degrees).[8] In 2001, however, after choosing to invest heavily in building its highly successful fiber-optics business and resultant punishing decline in the stock price when the bottom fell out of that market, Corning's chief technology officer, Joe Miller, formed the Exploratory Markets and Technologies Group (EMTG). At the outset, that group was composed of two teams of three people each (there are more teams now). Each team had a marketing lead person, a marketing analyst, and a technical lead person, and each team was devoted to a different area of Corning's technical competency expertise. Their role was to visit other industries, companies, and governmental agencies with which Corning does not necessarily have a current business relationship and identify key parts of important, complex systems in industries that Corning could apply technical expertise toward to solve big problems. Each opportunity they identified was written up as a white paper and reviewed by the internal Exploratory Markets and Technologies staff for promise as a real opportunity. Those that received positive reviews were then subjected to an intense market and technology assessment and value proposition development. These in-depth assessments were presented to the new technologies corporate oversight board, the Corporate Technology Council, for a decision regarding whether and how to move forward.

Mark Newhouse, vice president of strategic growth (a newly created position), labels these opportunities "keystone components": they are combinations of materials and process innovations that enable systems-level solutions to complex problems. This is the way that Corning has achieved business success throughout its history. For example, its competence in glass production, along with its development of the ribbon-cutting machine, enabled the production of the glass envelope for the light bulb, which triggered the mass market in light bulbs in 1926. The combination of germanium-doped silica (glass) development and chemical vapor deposition processes developed in 1970 formed the basis for optical fiber production. The fusion forming process in conjunction with the development of alkali-free glass created the glass substrates for liquid crystal displays that Corning now supplies to the world market. Every success that Corning has enjoyed has been based on these keystone components of difficult system level problems.

The Activities That Comprise Discovery

These two examples show that discovery takes many forms and plays out in many ways depending on the company context. But however it plays out and wherever it is located in the company, discovery as an organizational capability encompasses three categories of activity: foundational knowledge in multiple domains, opportunity generation, and opportunity articulation. Let's examine each of these in some detail.

Foundational Knowledge

Technical, scientific bench lab work is part of foundational knowledge, and it's the part most people think of when someone refers to discovery activities: materials characterizations, chemistry, physics, petri dishes, mathematical algorithms, feasibility testing.

Not all industries, however, depend on physical or life sciences as their basic disciplines. The creative industries, such as arts and entertainment, and the financial services industries (insurance, financial instruments, banking) require other sorts of foundational knowledge. Leonardo da Vinci produced many breakthroughs in portraiture because he was developing new foundational understanding of paint pigments and studied shadowing and light from a physics perspective. His deep knowledge across multiple disciplines enabled the discovery of how to create depth, facial expression, and overtones that his predecessors could barely imagine. The breakthroughs in combinatorial chemistry that are enabling new drug discovery processes in the pharmaceutical industry today are based on complex mathematical algorithms, simulations, and modeling. The industrial design industry, led by companies like IDEO, has developed breakthrough ideation processes based on the science of creativity and cognitive and social psychology as their foundational disciplines. Financial innovations such as hedge funds are based on deep understanding of probability, statistics, and optimization.

In most of the industrial companies we observed, some portion of the R&D budget is still reserved for fundamental research, and a larger portion is called directed research, in service of clearly articulated strategies for the company's future. Both of these parts of the R&D budget are differentiated from the large portion that goes toward servicing the ongoing, immediate needs of the business units. It may be the smaller portion of their budgets, but it helps enable breakthrough innovation. Services firms too have mechanisms for ensuring they're engaged. It may not be called R&D, but it happens. Investment banks are hothouses for economics professors on sabbatical, who find the stimulating environment an attractive place to go to explore complex mathematical problems that may someday serve a productive purpose.

In these days of open innovation, discovery also involves evaluation of externally generated discoveries and consideration of combinations of technologies from a variety of sources. Many companies place investments in small technology-based companies or take part in venture capital syndicates that invest in a portfolio of such start-ups, in order to maintain awareness of technology spaces that they do not wish to fully fund on their own. Many companies develop their own venture funds for investing in early-stage companies. In our main sample of twelve firms, seven placed such external investments.

Not only does an open innovation model provide a window into new technologies, it helps broaden the company's foundational knowledge beyond its core. This is imperative, since breakthrough opportunities frequently arise at the interfaces of technological or foundational domains. Ron Pierantozzi, director of corporate new business development working in Air Products' commercial development office, told his discovery staff that he wanted them to attend at least two conferences each year but that they should be different conferences every year. He did not want them hearing the same topics, seeing the same people, and even reading the same journals. He wanted them to broaden and deepen their knowledge bases constantly.

So industry by industry, the foundational knowledge disciplines may vary widely, but our point about discovery is this: for a firm to engage in breakthrough innovation, it must be actively involved in developing or engaging with communities that develop the foundational knowledge of disciplines related to their domains.

They also must be actively engaged in learning foundational knowledge in domains that could be combined with their own to create entire new domains in the future. Otherwise they will not sense opportunities until it's too late.

Opportunity Generation

The second of the three discovery activities is the work involved in finding potential business opportunities on the basis of technological discovery or other sources of ideas. It is not technical discovery in and of itself. In Kodak's System Concept Center (SCC), the alpha team was a group of creative people, some with technical backgrounds and others without, whose role it was to come up with business ideas for the company. This group had rather low turnover due to their creative nature and exposure to the history of idea generation within the SCC. They did not keep trying to reinvent the wheel.

A number of our companies used idea hunters to scout breakthrough ideas inside and outside the company for potential opportunities. At MeadWestvaco, for example, founders of start-up companies that the firm had acquired were assigned to the company's innovation hub. Their job was to work their external networks to find new business opportunities for the company. The reason they were used this way was that Laura Pingle and Rick Spedden, who ran the innovation hub, wanted to leverage the opportunity recognition skills of these proven entrepreneurs.

The GameChanger group at Shell Chemicals formed an externalization team devoted to developing future trend analyses based on visits to universities, and built a "hunters' network" of creative individuals throughout the company who'd submit ideas for new business opportunities that they found among their colleagues throughout the company.

At Sealed Air Corporation, the technology identification process team, composed of research directors and business development managers, is tasked with finding new opportunities to help fuel business growth extending five or more years in the future. 3M and DuPont, in addition to Corning, have formed exploratory marketing groups in their central R&D departments, charged with finding breakthrough opportunities at the technology-market nexus. One of our companies built an informal network of external contractors to generate and develop wild ideas and inventions. This network is maintained and funded by a senior executive who elected not to bring them into the company for fear that their creativity would be stifled. Having them outside the company, he believed, allows them to push back in terms of clarifying why their ideas are in fact opportunities, when insiders, content with the given norms in the company, would not. Idea cafés, idea jams, and innovation fairs are all mechanisms companies are using to surface opportunities.

The point is that companies are proactive, engaged with others, moving beyond the lab, investing time and money to generate new breakthrough opportunities. They're not waiting for a light bulb to go off in someone's head, or an Aha! experience. Different approaches maintain freshness and creativity and draw in different people, thereby maximizing the chances of identifying novel opportunities.

Project Approach or Platform Approach? One other point about opportunity generation: when most of the companies in our study started their innovation initiatives, they issued calls for proposals and thought in terms of projects. Within a couple of years, nearly every one of them had moved from a project approach to a bigger, broader, more strategically driven platform approach. Some were technology platforms, such as nanotechnology powders, coatings, and springs that can be used in a wide variety of applications, and some were business platforms, such as a decision to move into energy storage and delivery systems, for example, in which any technological solution is considered so long as it addresses the business problem of energy management. It almost doesn't matter.

In building a breakthrough innovation capability, projects aren't the way to go. Platforms create an increased number of options because they can be the foundation for a variety of business models, products, and applications. At the same time, they focus idea generation in domains of strategic interest for new business growth as opposed to one-off products. They accelerate the company's learning by leveraging new networks for multiple possibilities and building capabilities in a new technology or market domain that can be leveraged into many new product opportunities.

Analog Devices originally developed the accelerometer chip with the intention of using it to enable new opportunities in the automotive industry. The first application was as an airbag detonator, but Analog Devices ended up developing an entire accelerometer business, with applications in medical equipment, gyroscopes, and video game consoles. Platforms feed new businesses. Individual, discrete projects are, as one of our participating managers stated, "like pimples . . . annoying."

Opportunity Articulation

The third activity necessary for a discovery competency is opportunity articulation: the ability to clarify opportunities in a manner that energizes management. Business opportunities can be identified or generated, but if a story cannot be woven about their linkages to the company's future, renewal, or strategic intent, they don't go far.

Of course, to do this, the company's leadership has to have conversations about its strategic intent—not its current strategy, but where it sees itself as an organization in the next ten years. It needs to ask: What technology and market domains do we see emerging that we should be participating in or dominating? What new markets should we be investing in creating? What are we doing now to make that future happen? We're amazed at how few companies ask themselves these questions.

Strategic intent sets the framework or boundary conditions within which opportunity identification activities are framed.

It sets the boundaries for what is potentially acceptable and what is not. For this reason, the importance of strategic intent becomes critical to effective discovery activities.

Companies develop strategic plans on an annual basis, but many of those plans are extrapolations of the present. When that is the case, every articulated opportunity will be evaluated in terms of what it can add to the current organization, not the organization of the future. That means that many breakthrough ideas will be rejected because they don't fit or they are accepted without a clear sense as to why and then abandoned later. As one R&D vice president told us, the breakthrough evaluation board at his company had trouble discerning a good opportunity from a poor one. "When you have no sense of where you're going, every path looks equally appropriate," he told us. So in that company, project proposals were funded on the basis of the champion's passion rather than on something else. How strategic is that?

The more linkages the opportunity has with various dimensions of the company's strategic intent, the better. Why? Because priorities in objectives seem to change. As John Wolpert—who ran IBM's program for developing young talent, technology, and business innovation named Extreme Blue—told us, when one priority comes off the table, another one seems to take precedence. To the extent that the program supports multiple objectives, the more likely it is to remain in favor. Understanding this natural force in companies is a real help when working to articulate opportunities that could, and likely will, change the game for the company.

The incubator director in the advanced materials business at Air Products, Larry Thomas, noted:

> I don't think any BI program is successful in a corporate environment unless it lines up with multiple strategic and tactical models. Our battery program lined up with a research directive to investigate a particular class of molecules; with a corporate imperative to identify new growth markets for the company; with a corporate

strategic focus on energy; with a performance materials divisional desire to find new value-added materials that could serve as the foundation for new strategic business units; with an electronics divisional effort to find other ways to leverage our fluorine manufacturing and process infrastructure.

No one (or even two or three) of those would have been enough to gain support for the program, but the convergence of all of them was enough to make this program stand out as an ideal program for the corporation to flood with resources to see where it could go. Otherwise, it would have been more likely to languish as a half-a-man-year technical effort with a little bit of commercial oversight from your or my commercial organization, and we never would have had the resources to get past the substantial technical challenges associated with bringing a new material to market against an established incumbent, or explored the larger opportunities for the corporation in battery materials regardless of the success of the initial offering.

The model in my head is the syzygy event in 1982, when all of the planets were within a narrow arc on the same side of the sun. They don't all have to be in a line (like in the opening credits of *2001: A Space Odyssey*), but they all have to be pulling in generally the same direction.

Opportunity articulation takes work. It involves networking with markets to discern how this opportunity might play out as a business for the firm. What is the business concept? (Not the technical discovery . . . but the reason anyone in the market would care?) How big? How robust? How many applications? What are they? How would the business unfold, in terms of first application and follow on markets? In which aspects of the opportunity does the market perceive value? What are some alternative business models? And, most important, what will happen if we do NOT invest in this? Threats get much more attention than opportunities.

In one of our companies, the VP of R&D brought an opportunity before the senior executive council and it was soundly rejected. Why? He told us that he hadn't articulated it with strategic clarity, in terms of what it could do for the business. Rather, he'd described it as a technical enthusiast. He realized the mistake immediately, but it was too late.

Opportunities have to be articulated to multiple audiences in the company. The first is the evaluation board that must fund it. Usually there is some seed funding for nascent opportunities. But when it is a new business that will eventually reside in a business unit, that unit's leadership must be made aware. Michele Nelson, head of business development in 3M's R&D organization, told us that much of her job is opportunity translation for business units. "I tell them, 'We understand what your business is, and this is how this technology would enable you to go after these customers.' We help put the business translation link between the technology and the divisions so they can see what the opportunity is, the general scope and size of it, and then when they have that information, they can begin to incorporate it into their plans."

The Objective of the Discovery Capability

Discovery is a highly multidisciplinary competency that requires all sorts of technical backgrounds, as well as entrepreneurial, strategic, and marketing savvy. Most companies have migrated from technology identification activities in R&D to also incorporating business identification activities. This does not mean that R&D scientists need to be the ones doing all of the business-related activities. We have seen this expectation in some companies, but finally, more and more companies are realizing that new roles must emerge within the technical function—the new business creation sorts of roles like Michele Nelson's business development role at 3M, the exploratory marketing teams at Corning and inbound marketing teams at DuPont, the Systems Concept Center folks at Kodak, and the business development roles at Air Products. All of these people report, directly or through their immediate superiors, to the chief technology officer of their companies. They may have had their original training in a technical field, but they are not doing science or engineering. They're doing opportunity genera- tion and articulation—the beginnings of new business creation.

The objective of the discovery capability is to identify new business opportunities with game-changing potential that will set the course for the company's long term ever-evolving identity. These are opportunities that will form the company's future. The company will become them. What is needed is a management sys- tem to ensure accomplishment of these activities.

The Discovery Management System

Breakthrough innovation needs its own management system, the elements of which are reprised in Figure 3.1. We alluded to the idea that for each of the three building blocks (discovery, incu- bation, and acceleration), these system elements will play out a

Figure 3.1 Elements of the Innovation Management System

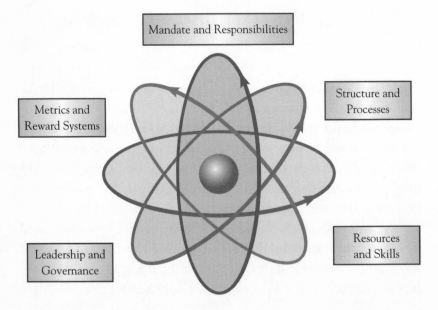

bit differently, because each building block is trying to accomplish something different in the overall system. Here we look at how it all plays out for discovery.

Mandate and Responsibilities

The purpose or objective of discovery is clear: it is the engine of breakthrough innovation opportunity generation. Only a fraction may ultimately become new businesses for the company, but they're all opportunity spaces that could potentially bring major new benefits to the market in some way or another.

Mandate is a bigger issue. The mandate for the breakthrough innovation function as a whole—the discovery, incubation, and acceleration system—can be expressed along any or all of these dimensions: degree of alignment with current businesses, time horizon, and domains. Table 3.1 presents mandate statements for each combination of time horizon (near term versus future) and alignment. While this seems inordinately detailed, it affects the BI function's relationships with other groups in the company. Business units may have new business development groups or their own marketing and product development groups that are searching for opportunities. Although the constraints levied on the business units regarding short-term profitability delivery are typically so tight that they cannot have the freedom to experiment with breakthrough opportunities, clarity of the mandate for each group is critical to ensuring cooperation over time.

By *degree of alignment* we mean the extent to which the BI function is expected to generate new businesses that fit or stretch the company's current lines of business or current structure. An aligned opportunity fits the current structure—or cannibalizes it. Aligned breakthroughs typically have a cannibalistic nature to them. Examples are Kodak's move from film to digital technology, completely disrupting the Consumer Imaging division's business model, or GE's migration from conventional to digital X-ray technology in the GE Medical Systems (GEMS) division.

Table 3.1 Potential Mandates for the Breakthrough Innovation Function

Time Horizon	Degree of Alignment with Core Business		
	Aligned or Stretch (Adjacencies)	Unaligned (White Space)	Multialigned (Gray Space)
Now	Find adjacencies, or closely related opportunities, that we can capitalize on now on the basis of applying current know-how into new spaces or bringing novel technologies to current markets	Find combinations of our current competencies and market networks to generate new businesses immediately	Find business opportunities that many of our business units should contribute to and benefit from
Future	Provide business units with far-future opportunities	Influence strategic intent by identifying platforms and domains of expertise on the horizon that the company should be gaining competence and expertise in to formulate the next wave of businesses for the company	Identify businesses of the future that draw on and benefit multiple business units in our organization

A multialigned business combines or feeds multiple current lines of business, such as Kodak's OLED (organic light–emitting diodes) division, which was originally formed as a platform-based division that the others could draw from as well as one that sold to external customers. IBM's pervasive computing EBO was multialigned, in that it required resources from multiple divisions as well as benefited the customers of multiple divisions. Finally, businesses that operate in the white spaces may draw on some core competencies of the organization by stretching it into new domains, such as IBM's move into the life sciences arena based on its mainframe competency, but requiring the addition of expertise in computational math as well as pharmaceutical industry participation.

To the extent that the mandate addresses unaligned opportunities, those new domains or white spaces will need to be identified. Some of our companies articulate them in terms of technology domains ("We need to own the bioinformatics space in ten years"), some in terms of markets ("the market for personalized health information management"), and some as a combination ("We need to be the leader in providing space travel for entertainment and vacation purposes"). They should be specific enough to help direct the placement of opportunity investments, but broad enough to allow multiple platforms, approaches, and serendipitous opportunities to crop up.

Finally, there's time horizon. Some companies focus their breakthrough innovation group on creating the future for the company, some focus on nearer-term opportunities, and others focus the group on finding adjacencies that are noncore but when combined with the company's expertise could bring new value over time. This last was the case of Corning's ongoing search for keystone components.

The most important thing to note about the mandate for breakthrough innovation in general, which affects the screening and evaluation of discovery activities in particular, is that it gets confused over time for several reasons. First, discoveries, ideas, and opportunities cannot always be planned or directed. The initial ATPs at GE were clearly aligned with business units' future needs. However, as the ATP teams kept working, opportunities emerged that crossed business unit spaces or for which there were no obvious divisional homes. That happens, and it's a good thing. Sometimes that's how blockbuster businesses emerge. One has to be prepared to recognize them as opportunities and not ignore them. The link to strategic intent is important, but if it's too tight, there's no room for serendipity or opportunism. The discovery function should influence the strategic intent of the firm as much as the strategic intent of the firm guides discovery. Reciprocal influence is critical.

Ron Pierantozzi's new business development group at Air Products identified an opportunity based on some scientific work done in its labs that would dramatically improve the storage capacity of lithium batteries and at lower temperatures than is currently available. This was certainly not something Air Products had thought about, but Pierantozzi's colleague Larry Thomas notes that he helped articulate that project in a manner that it fit within a number of Air Products' objectives for the future, in a force-fit sort of way. Ultimately the program's success has forced a strategic discussion about a new business domain that wouldn't have otherwise occurred.

Second, other BI groups we've studied found that although their original mandate was to focus on white space opportunities of the future, the day-to-day pressure to show progress caused them to select projects more aligned with the needs of the business unit. We have called this "mandate creep," indicating that a BI mandate can become compromised by requirements for short-term revenues. When that happens, a reassessment of the mandate should occur because aligned opportunities may be the responsibility of a new business development group, for example, in a business unit. If the pipeline's not rich enough to justify the innovation organization because white space opportunities are rare, perhaps the scope of the mandate should be expanded, but this should be done in conjunction with others in the company.

Our point is that the mandate seems to evolve and should be revisited from time to time. It may begin to overlap with others' mandates for innovation. And it's not clear how rigid it ought to be. If the BI group ends up dropping opportunities that aren't within its scope, it's a problem. But if, throughout the company's entire innovation system, there's a place for each sort of opportunity listed in each cell in Table 3.1, then quick hand-offs are in order. At one point during our study, Kodak had a clear innovation system in place. The Systems Concept Center (SCC) was supposed to handle white space (unaligned) and multialigned opportunities, both near and far term. Each business unit was creating a new business development (NBD) function that handled aligned but far-future opportunities. Each business unit had a traditional new product development (NPD) process that it used for immediate aligned opportunities. When the SCC generated opportunities that were aligned with business needs, either near or far term, it could hand those off to the NBD groups or NPD process teams.

At IBM, the multialigned opportunities are designated corporate-level EBOs, and those that are aligned with only a single division are divisional EBOs, nurtured and attended to by divisional rather than corporate leadership. This level of rationality, however, doesn't last long. People move and mandates evolve. That's why, every once in awhile, the innovation system has to be rationalized. What is the innovation mandate for the business units, and what is the innovation mandate for the corporate breakthrough innovation function?

The other elements of the management system should be aligned with the chosen mandate. We address each of the remaining elements here, and in Table 3.2 we summarize how they may differ depending on the mandate for the company's BI function.

Structure and Processes

The discovery process should be centralized, so that there is a known place to go with ideas and a clearly identified group whose job is to stimulate ideas, search for emerging trends and opportunities, and develop or actively engage in foundational knowledge. This group helps develop and articulate opportunities as potential businesses. The group should be tightly networked to the R&D communities in the company, and all company members should be aware of it so they can feed opportunities that may arise to the discovery group. Divisions of very large companies may establish their own discovery functions for aligned breakthrough opportunities.

**Table 3.2 Management System Elements Aligned
with System Objectives**

System Objectives	Aligned/Nearer Term	Unaligned (White or Gray)/ Far Future
Structure and processes	Structure: Report to CTO and maintain strategic relationships with business units. Processes: Planning with business units. Opportunity generation and articulation all oriented toward business unit strategic intent.	Structure: Report to corporate officers directly. Could be connected with R&D but not necessary. Processes: External network development critical.
Resources and skills	Resources: Corporate funded unless discovery work is done within the division. Skills: Networking internally is critical. Need novel technologies to bring newness to the space, since innovation occurs at the interfaces. How far afield can the business unit accommodate?	Resources: Corporate. Skills: May need technical skills not currently in-house. Need new business creation acumen, ability to reach into domains not currently connected to.
Leadership and governance	Culture of creativity plus discipline. Incorporate business's senior leadership along with senior technical leadership to evaluate opportunities.	Culture of creativity plus discipline. Incorporates corporate strategy, CEO, highest levels of corporate leadership along with CTO.
Metrics and reward systems	Number of programs approved. Number of programs adopted by business unit.	Number of new platforms identified and proposed, robustness of those, and impact on company's strategic intent.

More than one of our participating companies noted to us that several of their divisions are larger than the vast majority of companies worldwide.

Processes for discovery are fairly well known: building readiness for opportunity recognition through conducting scientific and technical work, engaging with internal and external technical communities, and holding futures workshops. Scanning externally for megatrends through the venture capital community, universities, and technical conferences helps the discovery function identify growth platforms for the future and influence the development of the company's strategic intent.

Discovery processes for opportunity generation, elaboration, and articulation include all of the follow-on work to check out or evolve an idea into a business possibility. Early market interaction to clarify the potential value and the extension of the current network to incorporate new market domains helps develop a rich list of potential applications. Opportunity articulation defines potential business concepts, possible business models, and links to the strategic intent of the company. All of these possibilities will be tested in incubation, but in discovery, that's all they are: possibilities.

Resources and Skills

Corporate resources are required for discovery activities that address breakthrough mandates that extend beyond any single business unit's interests, or even those that are aligned with the unit's but whose time frames are longer than the business units are accountable for. Some companies hold their divisions responsible for their own far futures and reserve corporate resources for white space and multialigned opportunities, but most do not.

A number of skills are important for discovery. Scientific prowess is needed to create or evaluate novel technology or combinations of technology and foundational knowledge or to enable learning in other domains. Market vision can articulate the opportunity in a manner that clarifies its value to potential internal constituents, customers, and value network partners, and strategic insight is needed to clarify how the opportunity enhances the firm's position in relation to its future objectives. These skills need not all be present in one person. It takes several roles to completely staff a successful discovery function.

Skills and attitudes common to all discovery personnel include a willingness to try many approaches and endure failures and an ability to endure rejection when an opportunity one favors is not selected. There's a high need for people with the ability to restart and look at problems in many different ways. A lack of bias is important. One of our BI discovery leaders was perplexed about how to populate her teams: "If I put a physicist on it, it'll be a physics problem. If I put a chemist on it, it'll be a chemistry problem." Gary Einhaus at Kodak made a similar statement when he said he needed "technically neutral" teams.

Leadership and Governance

In most industrial firms, the chief technology officer typically is responsible for discovery at the highest levels but needs strong partnership with the chief strategic officer to ensure influence on corporate strategic intent. The culture of discovery must be one of cultivating possibilities and focusing on big hits through exploration and imagination. The discipline and rigor associated with opportunity articulation attenuates the "wild idea" culture, however. Discovery leaders walk a fine line between energizing people to consider limitless possibilities and requiring a rationale, based on early market checking, to bring some level of business confidence to the discovery activity.

Cultural differences in firms' orientations to risk can change approaches to discovery. Some companies, like GE, will not take no for an answer. They identify big platforms or domains, develop competencies within them, and mine them for business opportunities. But the platforms themselves remained fairly stable once Scott Donnelly set them up. Sure, they add one once in awhile, and one, in fact, was defunded because it turned out not to be big enough. But over the four years that we tracked them, they stayed relatively stable. GE is risking lots of money, in some ways, by taking this "all eggs in a few baskets" approach. Nevertheless, this determination to build businesses in these spaces means that not every project within a platform will succeed, but by and large, a business will be built on the basis of that platform. A portfolio is generated within the platform. It's a portfolio-within-a-portfolio approach. It's quite calculated, and the culture and leadership surrounding these are strategically oriented.

Other companies prefer to place small bets and see which ones gain traction. This is more of an opportunistic approach to discovery, and the culture must lend itself to opportunism. Shell Chemicals' strategic innovation program does this. Anyone can propose an idea, and it's likely to get some funding for early experimentation. 3M operated that way for years and still maintains this approach to a degree. Most of these projects die off, but those that survive start to get attention. The company hasn't lost much in funding the early work. But it has a big investment in making sure ideas are coming in, get a good hearing, and are evaluated fairly. All companies that follow this approach are working to develop a culture of opportunism.

Still other companies don't like to invest in the early work at all. They'll invest in university research centers or small external companies, and when the technology development is coming along and appears promising, they'll bring it in-house. We call this a risk aversion approach. The sources of technology are external to the company and will not be bogged down inside it.

Deciding which opportunities should make the transition to incubation once they've been developed and articulated is a key function of discovery leadership. The criteria for such a decision are quite simple, since so little is known at this point about each project. First, can we demonstrate technical feasibility? Not under all conditions, but at least in the lab, or theoretically, with a planned path forward for further development? Second, is the opportunity big enough and robust enough? Could it be a line of business with multiple product or service lines or market segments? Can a near-term and then follow-on applications be envisioned? The reason we look for robustness, meaning many business options, is that promising avenues can dissolve into thin air when you start to probe them. We do not mean that you have to feel highly confident in any one of these; it would be too soon. That's the purpose of incubation: to test them all. But there should be a number of options at the outset that can be listed. Finally, does it help our company fulfill our vision of what we want to be in the future? Even if it doesn't fit right now, that's okay, so long as it's taking us to an articulated future. Companies, like people, need to dream. It helps to discern which opportunities are worthwhile and which are not important. They may all be attractive financially. More likely, none will appear financially attractive given the risk and uncertainty associated with truly radical innovations. So the strategic intent question is a critically important discernment criterion.

Metrics and Reward Systems

How should performance at discovery be measured? Many companies tell us they are rewarded for the number of patents received. That may be a critical foundation for breakthrough innovation, to be sure, but there's more. The number of new platforms identified and adopted, technical progress on any of them, richness of the business concepts in terms of robustness, and longevity and number of potential product lines or applications that could be generated are all key output measures. But there's also learning. Has the discovery team improved its approach to scoping an opportunity? Has the company developed new technical competencies or engaged in new market or technical networks because of the discovery activity? Has the discovery team gained the trust of others in the company, so that they're sought out for help when someone has an idea that he or she doesn't know how to articulate? As the BI portfolio ages, is the discovery team continuing to refresh the pipeline? Is it interfacing well with the incubation group and business unit constituents for aligned opportunities? All of these activities are necessary for the success of a discovery system.

Those involved should be rewarded for their contribution to any or all of these. We've seen situations where inventors, or patent holders, are idolized in companies, and frankly are treated like celebrities. This promotes a culture of invention that may or may not be relevant to breakthrough innovation. While it's impossible to know if any creative insight or invention will result in a business (since there's so much uncertainty to be surmounted), the others in the discovery equation share equally in the task of discovering if the possibility is worth exploring further and should be so rewarded.

In sum, the discovery building block comprises that unique combination of management system elements that create a capability to find, investigate, and articulate business platforms of potentially breakthrough magnitude, in accordance with the company's charted course for the future.

Challenges in Building a Discovery Function

We've noted that mismatches exist between the desire for breakthrough innovation and reality in a number of companies. These conflicts have to be faced if a company expects to initiate breakthrough new businesses. We summarize those that are specific to the discovery building block. In addition, Appendix B at the end of the book provides some useful diagnostic questions for you to assess your company's discovery function:

• *Companies desire breakthrough innovation but do not have deep foundational knowledge or are not organized to leverage it.* One company responded to its new leadership's call for game-changing innovation by forming a BI group, developing learning-oriented processes, and hiring an outside consultant to help it identify domains for cultivating opportunities. The trouble is that the company's R&D community consisted of process engineers who had no connections to pockets of scientific expertise to learn about leading-edge technologies. It's hard to get breakthroughs on the basis of commonly held knowledge. A second company had a small research group, but it was dissociated from the innovation function. Thus, the invention engine was separated from the innovation engine.

- *Companies confuse breakthrough innovation with diversification efforts.* One company's BI group issued a call for proposals that would move it into the green (environmentally friendly) marketspace. Proposals came in that had to do with recycling materials, public relations campaigns, and other attempts to change the company's brand image or diversify it into businesses it was not currently in, but none of them were breakthroughs. Clarifying the mandate and ensuring that the group is focusing on breakthroughs (however a firm elects to define it) is important. You'll hire people and institute systems designed for managing highly uncertain, potentially complex projects. Diversification, while important in its own right, is a different beast, and should be managed differently.

- *Companies desire breakthrough innovation in businesses that they've been involved in for decades.* It's a bit tongue in cheek that we describe the potential for aligned, near-term breakthroughs. If those opportunities existed, wouldn't the company have already leveraged them? The action is at the boundaries of leading-edge technology, new knowledge, or combinations of technologies. That's where insights occur. Even Michael Dell's famous business model of customized PCs delivered by mail order required a unique combination of knowledge about logistics, computer hardware, operational efficiencies, and marketing to formulate that game-changing business model. If instead he'd focused on building a cheaper computer by ensuring the components were manufactured more efficiently (which is what everyone else was focusing on), he'd never have come up with the idea.

- *When you attempt to change the culture of the company to one that generates and articulates ideas for the future, you may find that the floodgates open and ideas come pouring in.* Discovery generates a wealth of opportunities, many of which the company will never invest in. Motivating people to continuously generate and articulate opportunities that do not see the light of day is a challenge. People need to be selected and rewarded appropriately. And you may find that they burn out and need to be replaced after some time.

- *Companies focus on discovery without considering incubation and acceleration.* Discovery is critical to an effective BI management system. Most companies start with a discovery building block, but many companies end with it too, and they wonder why innovation doesn't get done. So let's move on to Chapter Four and explore the next building block, indeed, the most challenging one: incubation.

Questions for You

1. What aspects of Discovery does your firm do well? What aspects are managed poorly? What aspects are missing?

2. Are there people resident in your company with opportunity identification and articulation skills?

3. What challenges does the addition of commercially oriented people to R&D create? What is happening in your firm to counterbalance those challenges?

4. How does discovery differ for aligned and unaligned opportunities? Which are easier to identify and raise in the organization?

5. What incentives exist for the discovery organization?

4

INCUBATION

The Long and Winding Road

How many of you have the patience to try a new application in two different markets, watch what happens and then modify the offering, try it in two more markets plus a revisit to one of the first ones tried, and then do this again and again? Each time technology development has to occur, requiring perhaps different resources from those on hand. All the while, you have to be considering how far afield the opportunity may be taking you from your original concept and what this does to the business's potential fit with the company. In fact, you do it until you figure out what exactly the market values in the offering, how to build a business around it, and how to create a growing customer base. This is incubation, the second of the three building blocks. Incubation challenges those of us who like closure and clarity and also like to know what to expect for any given action. It's hard to know what to expect when you're blazing new trails.

You may be wondering, "Isn't discovery about blazing new trails?" Well, yes, but it's only the beginning. Getting ideas is the

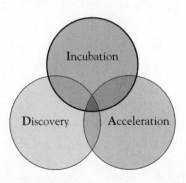

easy part. Developing those ideas into workable, strategically relevant business platforms is the hard part.

We describe incubation as a long and winding road because it takes the most time and is the riskiest of the three building blocks. It's also the capability that is most underdeveloped in companies today. Companies don't provide enough resources to incubation due to their orientation toward fast decision making and instant execution. As one project team member told us, "We've invested over $60 million inventing this technology. . . . I don't want to spend another $60 million inventing a business too!" Needless to say, we were shaking our heads in disbelief. But that's the reality. Companies do not realize that breakthrough technologies do not yield breakthrough businesses without enormous investment far beyond the technology itself, requiring lots of experimentation on many fronts.

The inbound marketing teams at DuPont were set up within central R&D to reach outside the firm's boundaries and find opportunities to apply DuPont's tremendous technical depth to problems in a wide variety of industries, as we noted in Chapter Three. Dick Bingham ran the APEX process that managed DuPont's portfolio of R&D projects. He also oversaw inbound marketing at the time and recognized the need for more involvement of that group in the APEX projects, as he witnessed a churn rate in the portfolio that he thought was too high. He was concerned that they may have been killing off nonobvious business opportunities because there was a lack of experimentation in terms of clarifying the business proposition in various application spaces. When projects took unexpected twists and turns, their choice was either to kill them or, if some degree of business acumen was resident on the team, test the new option. Dick told us, "For a while, we felt like if we did the initial opportunity definition, that was sufficient. But we're learning that you don't do a market analysis, voice of the customer, turn it over to the technical team, and expect everything to remain the same, or expect that you've really understood everything that you needed to in that first analysis. You need to stay with it . . . stay with the market connection continuously throughout the project."

Michael Blaustein, who preceded Bingham as the APEX portfolio manager but at the time was heading up another group, DuPont Ventures, was moving in the same direction with his portfolio of projects. In fact, the company decided in 2005 to create a marketing leadership development program and require a rotation through the inbound marketing group as part of that program. The idea was to develop the entrepreneurial side of DuPont's marketing talent in the company, as well as to ensure the APEX portfolio benefited from the additional business creation expertise that would now be available.

DuPont was a bit ahead of the curve of many of the companies we studied; it recognized the need for these resources but hadn't yet applied them. It hadn't confronted the challenges that companies face once they make these decisions: What sort of people do we need in these roles? And what do we need them to do?

A number of our company participants told us they were starting to get "marketing" help in R&D to nurture projects. When we asked if these were the right skills, we got long pauses over the phone. What, then, we asked, do you need? They couldn't say . . . ; it was just too foreign to them.

Defining Incubation

The objective of incubation is to nurture the portfolio of opportunities identified in discovery that have uncertain outcomes, but immense possibility for the market and the company. Many opportunities look promising, but as the due diligence takes place, most of them will wither. Those that do blossom may do so in unexpected directions and may even stimulate new opportunities that the discovery team may want to consider. That has happened in a number of our cases, most notably IBM's emerging business opportunities (EBOs). Several corporate EBOs have, through experimentation in the marketplace, stimulated brand new opportunities that have been the source of the next generation of horizon 3 initiatives.

Incubation is a business laboratory. It's the ability to experiment with technology, or discovery and business concepts and models simultaneously to arrive at, for any single project, a demonstrated model of a new business that brings breakthrough value to the market and, consequently, to the firm. Bear in mind that a good decision may entail no longer pursuing an opportunity because it's not making sufficient developmental progress or is evolving away from fitting within the company's boundaries for strategic intent. Marissa Mayer, Google's vice president for search products and user experience, notes that to do cutting-edge innovation, firms "have a theory of failing fast." Google's failure rate on its innovation portfolio is 60 to 70 percent.[1]

Incubation cultures therefore must make allowances for failures but hold expectations for continued pursuit of new frontiers. It's about the creation and pursuit of options, meaning continuing to invest some amount to stay in the game and investing small amounts in multiple directions simultaneously on any single project, so that when enough is learned, the company can move quickly to take advantage of the opportunity. The focus is on learning and then redirecting based on that learning. The ability to learn quickly is crucial. There's a constant focus on enriching and extending internal and external networks to enlarge the scope of the company's knowledge base and commercial opportunity space. Through these experiments, incubation helps spot new strategic growth frontiers for the company.

The incubation leadership team must constantly thin and enrich its portfolio, resulting in a fair amount of churn. Out of discovery come many opportunities, each with breakthrough promise. Those business concepts get tested a bit in incubation. Some dry up quickly, others redirect and spawn new options not originally envisioned, some take long periods of time to test from a technical standpoint and for the market to develop trust in, and others take off like wildfire. All the time, incubation leadership is culling, combining, and nurturing the opportunities; ultimately those few that are truly game changers are moved along to acceleration. Incubation may be somewhat expensive at the portfolio level, but acceleration requires serious investment in each business opportunity. The role of incubation is to cipher those opportunities that are truly new business platforms for the company, bringing wholly new benefits to the marketplace.

The companies we studied did not systematically practice incubation. Of the twelve participating firms in our primary study group, only one invested much in incubation when we first began observing in 2001. Over the next four years, nine of the remaining eleven companies recognized the need for this activity and attempted to build it. In our second set of companies, four of the nine had an incubation capability. Unfortunately, some of them diminished their funding of incubation over time as their breakthrough innovation mandates evolved toward projects that were aligned with current businesses (remember mandate creep, which we return to in Chapter Eight) or as the BI group experienced financial pressure to produce revenues increasingly quickly. Incubation appears to be the most fragile and least understood of the three competencies. That also makes it the greatest opportunity for rapid improvement.

Just as we distinguished discovery from R&D and from invention, we need to distinguish incubation from new business development and marketing. Many companies have a new business development function, and its responsibilities range from cultivating brand-new customers to managing the mergers and acquisitions activity for the company. New business development (NBD) can be about diversification without a focus on BI. Although both of those activities may be important for incubation, they are not the sum total of an incubation competency. There's a big difference between the experimentation activity we describe as incubation and the sales goals required of NBD people in most companies. The BI director at one of our companies distinguished the two this way: NBD is made up of marketing people, who work on an eighteen-month or shorter time horizon for developing new clients. Incubation in their industry may take three to five years. It involves defining a business model and requires deep technical development and interactions with markets.

Likewise, marketing is not the same as incubation. Marketing activities can range from market research to advertising, branding, and sales. It typically focuses on responding to current customers and fulfilling short-term objectives. Incubation, in contrast, takes a proactive (not responsive), hands-on approach to creating new market spaces by creating value through novel technologies, delivery systems, or other discoveries that may never before have been possible. It involves envisioning value chains that may not exist today and convincing other parties to participate. It involves testing and retesting perceived value in multiple market spaces to develop the strategic path forward for the business, its structure, and its strategy. Incubation is about new business creation.

Table 4.1 distinguishes the sort of market-related activities that take place during incubation (market exploration) from

Table 4.1 Comparison of Marketing Activities for Incubation and New Product Development

Incubation	New Product Development
Market exploration	Market exploitation
Design market learning to maximize variation to gain a view of the landscape	Design market learning to limit variety and gain deep experience with target group
Flexibility to learn how value is perceived across multiple markets	Refinement of positioning strategy in relation to competitive offerings
Experimentation	Execution
Focus: New connections, new potential partners, new application arenas to maximize options	Focus: Build loyalty, capture market share, saturate the identified market, leverage knowledge of current customers to maximize profits

those that take place in conventional new product development, where customers and technologies are, by comparison, better understood.

In this chapter, we discuss incubation at two levels. The first is the individual project level: How does one "incubate" a business? The second is the corporate level: How does one build an incubation capability to manage a portfolio of projects that are all being incubated? How do the BI management system elements play out for incubation? Let's look at individual projects first.

Incubating Individual Opportunities

At the project level, incubation is the work required to test any single business concept that has emerged from discovery activity. Although we listed as many possible applications as we could when we articulated the opportunity in discovery, the generation of new applications to enrich the business platform should occur continuously and will expand for platforms that truly are breakthrough as incubation-based experiments take place. We want to expand the original discovery to encompass as many possible uses at the outset as we can think of, so that we can experiment on them all because the original identified application might not be the best one to pursue in the short term.

When G. D. Searle first identified aspartame, the substance that would ultimately become NutraSweet, in 1965, it realized that aspartame could potentially be used in many ways, including chewing gum, ice cream, carbonated beverages, powdered drinks, cereals, whipped topping, spoon- for-spoon substitute for sugar, packets of sugar concentrate, and table sweetener. Searle tried them all, but technical difficulties abounded with many of these applications. For example, when sugar was removed from cereal to be replaced with aspartame, the cereal lost its bulk and became mushy immediately after it was doused with milk. Ultimately, however, most of these were worked out and were ultimately approved by the Food and Drug Administration in the mid-1980s (except the cereal).

It ended up taking twenty years for the technical, regulatory, and market uncertainties to be resolved and the business to be developed.[2]

A Methodology for Breakthrough Innovation Project Management: The Learning Plan

In breakthrough innovation projects, the shape of the ultimate market is usually unclear; which applications will gain market acceptance most quickly and fully is generally unknown, and the path forward is, in many cases, difficult to visualize.[3] In short, the severity and number of uncertainties make it difficult to define milestones and the pathways to achieving them for breakthrough projects. It is more reasonable and useful to identify and prioritize uncertainties that must be resolved, define alternative approaches to exploring them, and continually assess the value of cumulative learning compared to the costs incurred. Given this, a traditional approach to project management is untenable for incubating high-uncertainty breakthrough opportunities. Rather, a more learning-based approach is warranted.

A learning plan explicitly recognizes uncertainties on four dimensions that have surfaced over and over again as critical to manage in breakthrough projects we've studied:

- *Technical uncertainties* relate to the completeness and correctness of the underlying scientific knowledge, the extent to which the technical specifications of the product can be implemented, the reliability of the manufacturing processes, and maintainability, among others.

- *Market uncertainties* include the degree to which customer needs and wants are clear and well understood, the extent to which conventional forms of interaction between the customer and the product can be used, the appropriateness of conventional methods of sales and distribution and revenue models, and the project team's understanding of the relationship of the breakthrough innovation to competitors' products.

• *Organizational uncertainties* are inevitable given the length of the breakthrough innovation life cycle (often ten years or more) that organizational dynamism creates. All of the projects we tracked in phase I of our research had to contend with uncertainties related to organizational issues within the project and between the project and its various internal and external constituencies. The latter included organizational resistance, lack of continuity and persistence, inconsistency in expectations and metrics, changes in internal and external partners, and changes in strategic commitment. The uncertainties related to organizational context stemmed from a fundamental conflict between the mainstream organization and the unit engaged in breakthrough innovation, the difficulty of managing the relationship between them, and the challenge of managing the transition from breakthrough innovation project to operating entity.

• *Resource uncertainties* comprise the fourth category of uncertainty, as project teams continually struggle to attract the resources they require. For nine of the twelve projects in our original study, external financing made the difference between project continuation and cancellation. Resources in our conceptualization include not only financial resources but also competencies. In all projects save one, the companies we studied lacked one or more competencies critical to the successful pursuit of their respective opportunities. As a result, project teams, and especially their champions, spent extraordinary amounts of time dealing with resource and competency acquisition through a variety of internal and external partners.

Even when a breakthrough project is formally established, its funding and support are generally unstable over time. Interest in the project waxes and wanes as decision makers and sponsors come and go. Because the breakthrough innovation life cycle typically lasts a decade or longer, a project can expect to see its supporters and sources of funds change multiple times. Consequently project champions must be prepared to continually pursue funding from a variety of potential sources.

The experience at companies that are using learning-based approaches to project management is that generally the market uncertainties in the early stages of a project result in an overly optimistic assessment of the market opportunity. This can result in overcommitment of resources at an early stage and make redirection or elimination of the project more difficult. The learning plan methodology enables companies to address this issue and commit resources more appropriately during start-up and the front end of breakthrough innovation projects, thereby reducing the cost of project failure.

The heart of the learning plan methodology is the Learning Plan© Template displayed in Exhibit 4.1. The Learning Plan© Template offers a way for project teams to identify and catalogue uncertainties, the first step in confronting the chaos of breakthrough innovation projects. Briefly, the process of using the Learning Plan© Template is as follows:

1. The first step in using a learning plan is for the project team, ideally with facilitation of external experts, to systematically examine each of the four categories of uncertainty (technical, market, organizational, and resource), and record what they know, and don't know in order to identify holes in their knowledge that must be filled. Teams frequently record issues as known, but if they are challenged, they realize that these are deeply held assumptions rather than facts. Ideally this initial step of cataloguing knowns and unknowns will be as thorough as possible, thereby revealing as many latent uncertainties as possible.

2. After the cataloguing of uncertainties has been completed, the members of the project team rate uncertainties with respect to criticality as a basis for making decisions about the allocation of time and resources and prioritization of tasks. Criticality is the extent to which an uncertainty, if left unaddressed, could be a

Exhibit 4.1 Learning Plan© Template

Learning Plan Process Steps	Technical Uncertainties	Market Uncertainties	Organizational Uncertainties	Resource Uncertainties
Conduct learning loop:				
Define what is known and what is unknown in each category.				
Assess level of criticality (high, medium, low).				
Develop assumptions for each uncertainty.				
Identify, explore, and assess potential alternative approaches to testing each assumption.				
Select alternative testing approaches deemed most efficient in terms of learning per dollar spent per time.				
Establish measurement criteria for proving or disproving the assumptions.				
Define tasks and timetable for each test.				

(Continued)

Exhibit 4.1 (Continued)

Learning Plan Process Steps	Technical Uncertainties	Market Uncertainties	Organizational Uncertainties	Resource Uncertainties
Conduct the tests:				
Evaluate learning:				
Posttest, analyze, and assess what has been learned. For example, can an assumption be converted into a fact, or have we disproved the assumption? If the latter, what is our new assumption about the uncertainty?				
Explore how the learning affects assumptions about uncertainties in other categories (technology, market, organization, resources).				
Determine how the learning affects overall project progress.				
Define next steps required for subsequent iterations.				
Proceed with the next learning loop.				

Source: Rice, M. P., O'Connor, G. C., and Pierantozzi, R. (2008, Winter). Implementing a learning plan to counter project uncertainty. *Sloan Management Review.*

showstopper for the project. Low criticality uncertainties can be resolved in the normal course of the project, but those of highest criticality set the agenda for the project team's work. They are to be addressed first. All four categories of uncertainty—technology, market, resource, and organization—should be considered. What's interesting is that project teams may normally gravitate toward resolving technical issues since they may be most comfortable with those. But in many cases, it's organizational issues that may be most critical at the outset.

3. For each critical uncertainty, the team should articulate what it has been assuming implicitly regarding that unknown. What are the working hypotheses? These are the assumptions that will be tested.

4. Next, the team should consider multiple approaches for testing each assumption and select the tests to conduct on the basis of efficiency of learning, that is, how much learning is gained per dollar spent during the test period for each assumption test. Too often technical people engage in extended, comprehensive pure tests, or market research people conduct exhaustive surveys, when in fact the bulk of the learning might occur by visiting one or two key potential customers or engaging in a simulation activity. It is also important to select assumptions testing approaches that will satisfy managers with whom the project team must communicate.

5. The team and those who will evaluate its progress reach agreement on the objectives for each test and their associated time frames, and how the success of the tests will be gauged. (We have observed teams evaluating the use of a prototype at a customer site without being clear ahead of time how they will assess the outcomes in order to resolve uncertainties.) Once clarity about the testing process and outcomes assessment has been established, the team proceeds with the tests. With active and ongoing monitoring, the team will be able to assess the degree to which each uncertainty has been reduced, convert high-latency uncertainties to low latency, and reprioritize uncertainty reduction activities.

The second part of the learning plan template is targeted at evaluating the learning. The results of the assumption tests can be used to ask these questions:

- Have we converted the assumption into knowledge (either confirmed or disconfirmed it)?
- How has this affected our understanding of that specific dimension of uncertainty?
- How has it affected our understanding of the other three dimensions (given that they're typically highly interrelated)?
- What should be next steps based on the insights from this learning loop?

In some cases, the learning derived from any single assumption test may set the project back because unexpected insights emerge and additional latent uncertainties are uncovered. The team can choose to redirect based on the new information (which could be a bigger opportunity yet) or shut down the project.

A learning plan therefore has these effects:

- Catalyzes a comprehensive assessment of technical, market, organizational, and resource uncertainties
- Spells out assumptions about each uncertainty
- Presents alternative approaches for testing each assumption
- Prioritizes the assumption testing tasks and defines steps for moving forward as quickly and as inexpensively as possible
- Resolves each critical uncertainty through experimentation and learning
- Aids in the training and development of innovation project managers
- Enables appropriate communication with senior management and hence effective evaluation

- Serves as a log of the project's history—not only guiding the project through subsequent learning loops but also to develop a database for future projects

One pass through the learning plan, testing one set of critical assumptions, is called a *learning loop*. At that point, an evaluation with the team's oversight board or project evaluation board should occur, during which the results are reviewed, the new assumptions are clarified, and the next tests identified. If the outcome of the assessment warrants further exploration, funding and resources should be tied to the plan for executing the next learning loop. All members of the team and the oversight committee need to be clear and in agreement regarding the assumptions to be tested and the testing approaches to be deployed in the next learning loop and the time frames involved.

This iterative learning loop approach allows managers to decide on an ongoing basis whether the cumulative learning is of sufficient value to warrant continuing the project. At some point, promising projects accumulate enough insight—that is, reduce uncertainties to an acceptable level—to permit the development of a more traditional project plan and the adoption of a stage–gate approach to project management.

All of this work is directed at the ultimate goal of incubation: developing the opportunity into a business proposition, that is, a working hypothesis about what the new business platform could enable in the market, what the market space and value chain would ultimately look like, what the production processes would be, and what the business model will entail, with appropriate economic analysis. The initial work is to enlarge the scope of the potential business as much as possible, creating many options. Then for each option, the value proposition for the market must be clarified. "You cannot always know if it'll make you a ton of money up front, but you can usually figure out if it's going to be a loser from the beginning," Mark Newhouse at Corning told us. So it was for aspartame-coated cereal.

The only way to do this is through early market participation. Establishing customer pull through codevelopment partnerships or beta customers is key.[4] When Sealed Air was developing its oxygen scavenging-technology, the company elected to focus on an application in its core markets first to get into the market and demonstrate the value of the technology. So it began with a food application for pasta packaging in a familiar market. But many other applications existed, much further afield from what Sealed Air was used to. In describing one in the data storage industry, Ron Cotterman, director of packaging research and analytical services, mentioned, "I don't think we would have seen that application in the early days because, one, it didn't exist at that point in time, but also it was in a market area that we really had no contacts or expertise in, so we probably would have assigned that a pretty low probability of success. But the fact that we were going after something shorter term already [the pasta package], which had business value even though it wasn't going to be the big hit, created the opportunity for us then to branch out into other things. This bought us time to continue to develop the technology and identify these other applications."

Incubation isn't complete until a business proposal (or, more likely, a number of proposals) based on the initial discovery has been tested and at least appears to be exciting or, alternatively, is shelved because each direction that is pursued dries up. In incubation, many avenues are explored, but few end up being accelerated by the firm. There are lots of dead ends.

Project-Level Skills and Resources

Incubation requires resources for the team to conduct prototyping exercises, interact with the marketplace, and seek guidance along the way. That's why DuPont decided to extend inbound marketing's responsibilities to ongoing projects. That's also why Nancy Sousa, head of Kodak's Systems Concept Center (SCC), realized she needed to reallocate her budget toward developing prototypes so that the SCC teams could be gaining market participation with the technology to provide feedback as soon as possible.

At Sealed Air, Ron Cotterman's Technology Identification Process (TIP) team began to take a more proactive role in overseeing the projects they commissioned to help them develop the business's strategy as they learned from their customer visits and technical development work. Cotterman told us, after having been part of the program for four years, "We've decided to take a more defined role relative to management of the projects with the idea being that if we don't, it will be easy for people to say, "I have to focus on my short term activities." So we decided that TIP will take a more active role of helping to work with these incubation teams regularly, not just quarterly, and we would review their progress against milestones and work on setting some new targets with them."

Experimentation skills are required for incubating projects. The basis of the learning plan is to test assumptions, identify what we don't know, and make headway on finding answers. The learning plan methodology requires that experiments be conducted not only on the technical front but simultaneously for market learning, market creation, and testing the match of the business proposition against the company's strategic intent. Implications of this approach are that organizations need the right kind of people on their BI teams and managing those teams. These should be people who are comfortable with uncertainty, can move into experimentation, and are willing to be flexible in pursuing opportunities as they emerge. We find few firms seeking these kinds of people. However, those with a stage-gate kind of mentality will be ill suited to BI kinds of incubation work. This becomes a human resource issue and a strategic issue as to how the organization will recruit, develop, and retain these people who are so critical to the success of BI activities.

Incubation at the Portfolio Level

Incubation extends beyond a methodology for managing any single breakthrough project to a full-fledged competency. We found four distinct sets of activities needed to develop and maintain a strong incubation competency. First, incubation leaders spend time legitimizing the activity of incubation to the whole company. Given that it is so foreign, establishing awareness of this activity's value throughout the firm is a critical issue and requires an investment of time and resources. Second, incubation leaders provide support for project teams so they can experiment: coaching teams on their learning plans and strategic direction, brokering resources and connections for them inside and outside the company, and nurturing them personally since BI projects frequently fail, and careers can be threatened in large companies if they're associated with too many failures. High-uncertainty project management isn't easy, so we find the need for lots of coaching and hand-holding. Third, incubator leaders work to develop their staff, who are rare birds in most large companies. Finally, incubation leaders must monitor the balance in their portfolios. They are constantly thinning and enriching the portfolio on the basis of the experimental learning taking place at the project level as well as evolution in the company's strategic intent, and considering the portfolio's balance and diversification on a variety of dimensions.

Legitimizing Incubation

Although many CEOs say they want breakthrough innovation, most do not understand what it takes. "In their heads, they know they need breakthroughs, but in their stomachs, it's a different story," one BI leader told us about the senior management team at his company. At one company, the BI orchestrator made a number of presentations to the senior leadership team to educate them about the importance of working at the platform level to develop whole new businesses rather than just focusing on individual projects. In dealing with the evaluation board for the BI portfolio, he regularly began using the term *incubation* to encourage this kind of thinking among the senior leaders. In another company, the BI leader publicly linked the importance of incubation to recent project successes. He also called in three academic experts well known for their research on breakthrough innovation management processes to speak to his team as well as the head of organizational development for the company. He then developed a synopsis of the three experts' remarks to propose a path forward for their ongoing incubation development to his boss.

At GE, the heads of the advanced technology programs (ATPs) were rewarded for identifying to the chief technology officer any early commercial opportunities that arose from their work and their market participation, whether or not those potential revenues were in line with the ultimate vision for what the new business would offer. At that point, the project team would remain intact and continue working on the longer-term objectives of the program, but a new project would be initiated under shorter-term programs and would be funded independently to commercialize that opportunity. The initial project team was rewarded for identifying these opportunities but was not distracted from the long-term objective by having to focus on the commercialization process. That was left to others with the right skills. The early and frequent interaction with the market that this creates is useful for the entire BI program. It builds credibility and legitimacy within the company as it begins to stimulate market interest and revenues early. GE's chief technology officer made sure to leverage these successes by connecting the early revenues and market excitement to the long-term investment the firm was making in any internal talks that he gave. In addition, the heads of the ATPs were encouraged to speak about their projects at professional conferences and to publish papers, helping to legitimize the firm's investment in long-term research and business experimentation. Any public statements about the ATPs have been made in the context of recognizing their long-term nature. Stating it publicly helps legitimize it internally. At Johnson & Johnson Consumer Products, external experts were brought in to train project teams and general managers on the learning-based processes associated with breakthrough innovation, providing external sources of legitimacy for incubation as an emerging discipline.

How do we know that devoting effort to legitimizing incubation is so important? We also saw companies that engaged in incubation itself, coached project teams, and helped them run experiments but did not promote the role of incubation outside their small group. Unfortunately, the BI groups in those companies have been disbanded. So you may be doing it, but if you cannot convince others of its importance, your effort is doomed.

Providing Project Team Support

BI teams need lots of support. We saw three types of support that were notable. The first was strategic coaching. Second, incubation leaders helped broker relationships between the team and key internal and external people that the team needed. Finally, teams required, and in some cases received, emotional support in the form of nurturing. Operating in high-uncertainty environments while your surroundings constantly sound the drumbeat of the importance of predictability can be difficult at times.

Strategic Coaching. Scientists get great ideas but cannot be expected to also have all the perspective, education, and experience necessary to develop a business. They may not even have the motivation. But even if they are motivated and even if they have a person who can help with marketing, having an external person to help clarify direction among the many possibilities is very important given that each situation is new and the possible paths are endless. It feels like chaos. "They seem lost in the market because the market isn't even there. They don't know the processes to use," one coach told us. Another said, "Helping teams develop clear strategy is like water torture. I plant seeds. I shape people's heads."

At IBM, the team of three senior leaders we described in Chapter Three work monthly with each EBO team leader and team members. These are not reviews that result in a go-or-kill decision. Rather, these senior coaches spend time helping the fledgling businesses clarify their strategies and improve their in-market execution. They talk markets, business models, economics, technology direction, financing options, partnerships, staffing, and organizational design. They constantly press on the issue of proving out the value proposition. These are huge investments of time on the part of IBM's senior leadership: sixty man-hours per month. But it's paying off. IBM's EBO program delivered over $2 billion in new business revenue in its fourth year off the ground and continues to build.

Coaching experts asks strategic questions to help teams identify the showstopper uncertainties they're facing and therefore conduct the most important learning first. They help teams focus on the most critical issues, even though team members may find those uncomfortable. Coaches help drive the teams to constantly consider the end game in terms of what the business's value proposition is and how it can create new market spaces. They help teams figure out how to work around barriers and help them get access to resources.

At the beginning of our study, seven of the twelve companies offered no coaching support at all. So although there was a corporate initiative to build a sustainable breakthrough innovation competency, the experiential resources required to help create it were not in place. In one company, the BI leader did not see a particular need for such coaching. She described her trust in the individual project leaders' abilities to find their way and viewed her role as enabling whatever was required as needs arose. Yet when we spoke with the project leaders, those operating in the most ambiguous arenas, where application markets were undefined and technical development paths unclear, they expressed high degrees of anxiety and a need to learn more about BI and incubation.

Over the course of our four year-study, most of the companies did use coaches, though not formally, and the coaches weren't always expert at new business creation, thereby muddying the incubation waters. In several companies, coaches were drawn from the ranks of the operating units and had no new business creation experience. Although all indicated a keen interest and some were exposed to limited external training regarding management processes in highly uncertain environments, they hadn't lived it. In two of the cases where this occurred, team members who were being coached by these folks told us that the processes being imposed on them by the coaches were too rigid. It appears that in some cases, processes are put in place to substitute for experience-based judgment, and their impact is to rigidify people's behavior and minimize the variation of potential actions at a time when that could hinder the development of the project to its ultimate full potential. One team member told us, "Real innovation is about relationship, chemistry, trust. We need guidelines, not processes."

The coach provides strategic guidance, helping teams chart their course through areas that project teams do not typically address: their project's role in the overall strategy of the company, organizational commitment, resource needs in terms of money and competencies, structuring new markets to leverage the innovation, and internal marketing of the project. In addition, the coach helps project leaders become comfortable with learning-based project management processes, tools, and methodologies and helps them set realistic milestones. They identify barriers in projects that could impede progress and have an impact on the overall portfolio, and they work with the project's advisory board and the portfolio governance team to devise strategies for overcoming them. The coaching role should become professionalized and should be considered a prestigious position for senior people who have built an extensive knowledge base and nurtured connections within and outside the company. It should be used as a training ground for junior new business creation specialists only if a senior person is available to be a mentor.

Brokering Relationships. Besides talking with the teams, project support involves talking with others inside and outside the company about the team's fledgling business, brokering new relationships for the team so they can gain access to new resources and knowledge. Connections to politically important players in the company are critical, and the incubation leadership team can help ensure those connections are made.

Nurturing: The Emotional Side of Breakthrough Innovation. Many dark alleyways and many dead ends are associated with BI projects. There are successes, to be sure, but mostly there's ambiguity. As projects proceed through incubation, their original scope changes based on the team's learning. What do I do next? Is our project support secure? How do I show progress? How do I get my hands on that specific resource? The CEO thinks the idea is great, but the business unit leadership isn't interested in funding it. How do I deal with this uncertainty and these mixed messages? Are we drifting away from a business model that'd be accepted by the company? One BI leader told us, "I have frequent contact with my team leaders because they, like everyone else, like to feel the connection with the organization, like to feel that they're making a difference, and they like feedback, constructive and otherwise. So I probably see five or six of them every day, whether it's casually or a call." One incubation leader held regular social get-togethers among her project team leaders to help them maintain a degree of confidence, share their experiences, and make sure they felt valued given the risks they were taking.

In addition, breakthrough innovation work attracts some individualistic sorts of people, and ensuring they're connecting to the company's strategy is important. Also, many BI project development cycles are much longer than traditional projects that team members have likely worked on in the past. It's easy to lose a sense of urgency or to get discouraged when building a business takes a long time. But when you're working on a long-cycle-time business opportunity in the midst of other colleagues who are working on a project for several months, celebrating its success and taking on the next challenge, the time horizon and the reward cycle seem longer still. So several of our leaders mentioned the need to deal with this more personal side of their project leaders and teams.

Developing Incubator Staff

Coaching doesn't always have to be done by senior leaders. Sometimes they are the least likely to provide good coaching. They may never have built a new business from the ground up or may not be talented in that manner. Developing a staff to serve as coaches is key.

Three years after instituting the EBO program, IBM is using people who were part of EBO teams who have "graduated" into mainstream businesses as coaches for new EBOs. And they're documenting their experiences and developing business practices based on those experiences. IBM is building an incubation competency in this way.

Most of our companies recognized they had a severe shortage of "incubation talent"—either coaches or a pool of potential project leaders they could draw on. They had difficulty in attracting and retaining this talent, primarily, they felt, because there were lots of risks and few rewards. This is a sad truth. There is essentially no career path for excellent team leaders or, even more, for coaches and new business creation staff. One company's BI leader told us he's "gaining tactical stock value" simply by hiring promising incubation talent from the outside. People know incubation talent when they see it but have a difficult time describing what they're looking for.

The incubation roles, either team leaders or coaches, are viewed by some companies as a development opportunity for helping promising young employees become trained for general management experience. If they can develop a small opportunity, the thinking goes, they can learn responsibilities for larger business. For technical people, a rotation as a new business creation coach helps them become more attuned to commercial concerns.

At Air Products, Ron Pierantozzi, director of new business development, found several people who "were born to be facilitators" and keeps them on his staff permanently or lends them out and brings them back. At Air Products, MeadWestvaco, and Shell Chemicals, coaches attend workshops to learn about breakthrough innovation project management processes. Others learn how to coach by working with those already experienced. Ultimately those experienced in coaching or leading several projects are then rotated out of the BI coaching staff to the business units, to be ready receivers of BI projects in those business units and make way for others to gain experience in coaching project teams. These are the proactive examples, in which incubation leaders are using the coaching role to expose many people to the world of high-uncertainty innovation, thereby educating the company and slowly, ever so slowly, attempting to change its culture.

In many companies, however, coaches do not get much help, because the incubation leader may not have the necessary expertise. So coaches learn by making mistakes. This can be costly, especially if the coaching role is considered a short developmental rotation on someone's career path. Companies must either train them quickly or keep people in coaching environments, since experience across a variety of projects or in developing a business is what makes a coach valuable. Otherwise they shoot from the hip in their advice and are not viewed as worth the team's time. They become another obstacle rather than a resource.

In one of the companies we studied, the "incubation manager" had never incubated a business before and had never even attended a workshop to learn about incubation. He did bring in one consultant to teach teams a specific project management process but spent no additional time with that person learning how to coach innovation teams in the use of the tool. He was viewed by his teams as someone whose purpose was to enforce company policy, and therefore they ignored him.

At another company, the incubator manager was brought in from the operations side of the business, having had tremendous success in driving down costs and increasing quality. Put into the incubator, however, it became obvious to us that that manager was ill suited for uncertainty, ambiguity, and long-time-frame projects. Her management style remained the same as if she was in operations. Within two years, she not only left the incubator center; she also left the company.

In some companies that are focused solely on developing breakthroughs that are aligned with current business units, the BI leader depends on representatives from the business units to help guide the teams. At GE, the number of business program managers—representatives from the business units who nose around R&D and the ATPs as part of that, help the teams consider appropriate external partners and customers, and develop the business plan—has multiplied. At another one of our participating companies, the head of the BI portfolio noted that the incubation staff had been directed by senior leadership to coach the teams for only a short time, before moving them to the business units, so as to maintain strict alignment with business unit requirements. The problem with this approach was that the business units did not know how to nurture these investments so early in their development, and most fell out of sight before too long. The lesson here is that coaching is a commitment of time, money, and skill that has to happen somewhere in the organization for the new business to be fully explored and take root.

We have one caution about BI coaches' responsibilities. We observed in several of our companies that the coaches sometimes get so involved in the content of the projects that they end up doing the project work themselves. It's important to distinguish between coaching and executing. If the learning and knowledge sharing does not occur across projects through the coaching vehicle, then each project team is essentially operating on its own, not benefiting from the cumulative experience of other teams as represented by the coaches' feedback and questions. Coaches obviously need to be involved with projects, but their value is in helping each team make its way through the confusion of unknown markets, rapidly developing technology, waffling organizational commitment, and uncertainty about resources. Their benefit is derived from their experience across multiple teams, and they should be reserved for that purpose. As one of our respondents told us, "We need to develop process coaches. They should have two or three projects at a time. They'll become the bees carrying the pollen between the learning agents of our innovation network."

Monitoring the Incubation Portfolio

The fourth and final activity that occupies the incubation leader is oversight and evaluation of the BI portfolio of projects. Companies use the language of innovation portfolios, but few have become sophisticated in managing their BIs in accordance with portfolio principles. They don't behave as if they're using portfolio management. Rather, they are managing a collection of projects.

We address how to evaluate the incubation portfolio, and indeed the entire portfolio's health, using portfolio-based principles in Chapter Six. Here we focus on the tricky issues of incubation portfolio oversight. By this we mean decision making regarding project progress across all projects within the portfolio. The questions most companies are confronted with relate to portfolio governance (who oversees evaluation and what their responsibilities are) and criteria used to assess project progress.

Portfolio Governance. Churn rates are high in incubation as learning for each project occurs and many hit dead ends or are redirected completely. How should this frothy mix of highly diverse projects be monitored? Who should be responsible for overseeing it? For providing strategic direction to the portfolio and projects? For deciding whether and how to continue each project? For granting resources for each project's next learning loop? For looking for ways to build synergies among the portfolio businesses where appropriate? For addressing the impact of each fledgling business on current businesses in terms of their potential to compete with or cannibalize them or to dramatically move them in a different direction? For leveraging the company's resources and networks by opening doors, making connections, and removing barriers on behalf of the portfolio businesses? These are the responsibilities of the incubation governance board.

Nine of the twelve firms in our primary set of companies had formal governance or review panels to evaluate project progress. The sheer volume of proposals influences the nature and composition of these panels. Projects entering the incubation stage are numerous, and their evaluation and supervision require significant amounts of time and energy. The evaluation boards in several of our companies were initially composed of senior-level executives, but these boards quickly realized that the project review time was too great, and so they repopulated the portfolio review boards with upper midlevel managers.

Although we advocate an incubation leader to ensure that the incubation function is alive and well and that a healthy pipeline of projects is being pursued at any time, project reviews should be handled by a board representing a variety of constituencies in the company, to ultimately aid the new business platform's assimilation into the company's ongoing operations. Kodak's Systems Concept Center was run by one leader, but incubation projects were reviewed regularly by a venture board, comprising the director and vice president–level representatives from each business unit, as well as the chief technology officer. A corporate officer should sit on the incubation governance board to ensure that business-unit-level representatives are looking beyond their own unit's requirements and are in fact considering the company of the future as they interact with the project teams. DuPont's APEX review board was similarly composed, as was Corning's Growth and Strategy Council, which oversaw projects that had matured beyond the Corporate Technology Council's purview.

Most of our companies that grew sophisticated enough to fund a number of projects adopted what we call a "portfolio-within-a-portfolio" approach. In Chapter Three, we described how companies evolve from a project to a platform orientation, in which the platforms are linked to their vision of the company's major businesses of the future. Each platform spawns a number of projects within it, so the platforms themselves comprise a portfolio, but within each platform is a portfolio of projects.

It's difficult for a single governance board to oversee all those projects; for example, if there are five platforms with five projects each . . . , that comes to twenty-five projects. In fact, several companies told us that they could not assemble a group of people with enough knowledge and experience to span that range of technology and market space. Kodak, for example, ultimately set up advisory teams for each fledgling business. Nortel Networks' Business Ventures Group and IBM each did this as well. Since each innovation business will likely be pushing the boundaries of a different market and technology space, it's critical to ensure that each has a set of advisers with the domain expertise necessary to support them.

Assessing Project Progress. What should be the decision criteria that governance boards use to allocate resources to portfolio businesses to continue their work or, alternatively, to stop? Fledgling businesses are testing the assumptions associated with uncertainties they articulate in their learning plans. Each learning loop may test several of the most critical assumptions across the technical, market, resource, and organizational spectrum. Since each project's learning path is determined based on the critical uncertainties it faces, there is no way to predetermine a set of decision criteria for assessing a learning loop. Progress is determined based on the team's learning.

Over time, of course, the governance board will continue projects that are gaining traction in one or multiple application spaces, evolving in a direction that is favorable from a strategic intent standpoint, making technical progress, identifying new domains to explore even if those initially identified are drying up, and throwing off complementary knowledge and network connections to other projects in the portfolio. If every critical assumption is being disproved, technology development is not progressing at all, multiple applications have been pursued but little market enthusiasm is generated, or the project is devolving into an incremental innovation, it should be shelved if it's meeting dead ends or transferred to an appropriate business unit for quick development and launch if it's proving to be an incremental innovation.

If the review meetings are handled as working sessions to examine learning that was described for the learning loop (rather than as formal project evaluations), teams should come to the conclusion themselves that there's nothing more to learn and request that the project be iced. One issue is that teams do not always have passion for a project and so may try to shut it down without investigating all avenues. Although passion cannot be forced, due diligence can, and the power of surfacing the unknowns in the learning-loop format is that it helps to catalogue all of the due diligence steps that need to be taken. But there are also implications for populating a team in that, to the greatest extent possible, those most interested in a project should be assigned to it.

Two of our companies maintain an inventory of potential opportunities they develop in discovery and hold them as inventory. They call it their "bench of opportunities." Members of the BI function are aware of what is on the bench, and they're constantly evaluating the current opportunity they're working on against those on the bench. In their project reviews, teams have been known to suggest that their project be shut down because of its comparatively lower level of excitement and potential than others on the bench. What a contrast to other companies we've studied in which teams are afraid to admit "failure" of their projects, and so try to put a positively biased spin on the review board presentations to defer having a project killed. The governance board must set a culture of open discussion and honesty in the review meetings to ensure that they're promoting the businesses with the greatest breakthrough potential.

Incubation Management System Elements

Just as we described the ways in which each of the elements of the innovation management system plays out for the discovery building block, so too do we need to review these for incubation. Many have been developed in detail throughout the chapter, so we'll summarize. Once again we reprise our management system figure (Figure 4.1).

Mandate and Responsibilities

The purpose of the incubation activity is to nurture a portfolio of fledgling business opportunities through early market participation, technical development, and business model experiments to clarify whether and how they can become new businesses of breakthrough magnitude that further the company's strategic intent.

Figure 4.1 Management System Elements

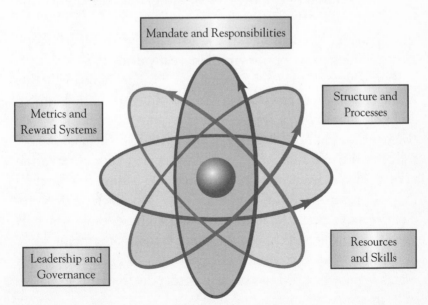

Mandate and Responsibilities

Structure and Processes

Metrics and Reward Systems

Resources and Skills

Leadership and Governance

Structure and Processes

In terms of an organizational structure and location for incubation, it's pretty clear that it should be tightly linked to, but not part of, R&D. Unaligned and multialigned opportunities need to be housed in a dedicated group under the corporate-level umbrella, since there is no fit with any single business unit. Location is more complicated for opportunities that appear to be aligned with a specific operating unit. If incubation occurs in the business unit (as many of our companies preferred), then the rest of the incubation management system elements must be in place to ensure the fledgling business is not extinguished or underexplored in the unit. Metrics, processes, and oversight must all align with the fundamental experimentation and variety-enhancing character of incubation. These characteristics conflict with operating units' mandates. Except in the case of IBM, where corporate officers personally coached the EBO teams and oversaw the funding allocations to them, even though the EBOs were resident in and resourced in the divisions, we have not seen a successfully incubated business when it is housed in the business unit. There's just too much pressure to conform to the culture of the present rather than the culture that forms the future.

Incubation processes are those associated with experimentation. Learning plans, options thinking, early market participation, and early harvests to generate some first revenues build credibility in the organization, teach about the cost structure of the business, and help identify new applications. Several of our participating companies, most notably 3M, began applying Six Sigma processes to breakthrough innovation over the course of our observation period. This is an oxymoron. The principle behind Six Sigma is to weed out error, reduce variability, and make each process predictable and repeatable. Innovation opportunities that ultimately upend industries and change the game in terms of bringing value to the market are too ambiguous at the outset to benefit from Six Sigma approaches. 3M has since backed away from this initiative in its innovation management system.[5]

Regular coaching sessions with teams are critical for helping them clarify the strategy for the business that is emerging as they probe the market. Having a backlog or bench inventory of opportunities helps teams honestly report their beliefs about the business's breakthrough potential.

Resources and Skills

Corporate resources (along with whatever external funds are co-opted by each fledgling business through codevelopment partnerships or government monies) are required for the long and winding road of incubation for unaligned and multialigned opportunities. Divisional resources should be used for aligned opportunities if business units are held accountable for their far future. Our participating companies believed it was critical for business units to share in the funding of aligned opportunities. If they don't have skin in the game, the theory went, they won't adopt the business when it's ready to move to the business unit.

Several incubator groups expected business unit funding but avoided business unit participation in incubating the young business opportunity. The idea was that business unit influence would pressure the fledgling business to conform to its own current customer needs, business models, and time horizons rather than allow experimentation on these dimensions to best understand the business opportunity's breakthrough potential. One company, in fact, used what it called a "forced adoption model." When an opportunity arose that appeared to be a potential fit with one of the business units, the BI group would inform that unit's senior leadership. But the unit was not allowed to participate on the oversight committee as the business was developing. Incubation of the business was funded by corporate funds. Once the business was to the point of annual sales in the multimillion-dollar range, the business unit was expected to take it over and reimburse the corporate incubator's expenses. The business had shed enough risks to the point that it could be accelerated within the business unit.

Incubation talent must be considered at two levels: project teams and incubation staff. Project team members are entrepreneurial sorts—opportunistic, flexible, and willing to experiment and adjust as they learn. They are synthesizers: able to take lots of different cues that do not seem to go together and plot a course of action from them. They are creative . . . , able to chart a direction without experience in that domain, since each business opportunity is unique and new. Their persistence helps them endure the long and winding road, and they must be resistant to closure so they can continue to generate and investigate options. Analytical skills must be resident on the team to develop economic models for the business, identify gaps in the business model, and consider mechanisms to persuade partners to participate.

Team leaders must have a high degree of interpersonal skill, since they must develop new networks for each project and must enjoy contacting people they may not already know (potential partners, customers, professional organizations). Strategic thinkers, they must constantly consider how the business opportunity will contribute to the company's future, and how the business might be resourced and run given the company's current structure.

Incubation coaching staff also must be able to learn quickly and help facilitate each business's progress through insightful questions and problem solving. They must be politically savvy to help broker relationships for each team and help the teams shape their opportunities in ways that are meaningful to corporate leadership. Incubation coaches are good listeners: empathic and nonjudgmental. They ask open-ended rather than leading questions to facilitate teams' learning and identification of next steps in their business's evolution. They know when to guide teams to diverge or converge in their thinking by recognizing patterns in the business's growth through their past experience.

Leadership and Governance

Firms should designate a senior leader in charge of incubation. Currently these activities typically report to the chief technology officer, who may have no business creation experience at all. That leader can set the tone for an incubation culture: one of new business creation rather than new product development. It's a culture of entrepreneurship and business building based on reaching out to new partners to clarify and validate business propositions. It's a culture of trial and error, learning by doing, and experimentation on many dimensions. Failure is an option; in fact, it's expected. Most incubating businesses will not make it to acceleration.

A governance board of appropriate business unit and corporate constituents should be used to review the incubating businesses. The criteria to use for determining whether a business is ready to move to acceleration is a demonstrated path to a profitable business based on breakthrough innovation. First revenues are coming in, a path to a number of follow-on opportunities appears clear, and there's reason to believe it'll be highly profitable. Partnerships have been established, and the business model that defines the relationships among those in the value chain is settled. Finally, there's a demonstration of commitment from senior leadership that this emerging business will help set the course for the company's evolution over the next generation.

Several of our companies told us that the governance board becomes impatient to show results and therefore move projects to acceleration before they're ready because the long and winding road is too long, and "everyone's watching." Although there need not be any hard and fast rules regarding the transition, businesses that are accelerated on the basis of very early but underdeveloped business proposals may be suboptimized. The first market may end up being the only market that the business leverages. (We offer more on this topic in Chapter Six.)

Metrics and Reward Systems

Metrics for incubation are all about learning, but also about lowering the business risks. Are they really breakthroughs worthy of a large investment? Is the incubation function having an impact on the company's continued development of its strategic intent, based on clarifying the new sources of value it can bring to the market? Incubation is functioning well if it is housing an appropriate number of opportunities given the company's capacity for innovation. In some companies, that may be two or three at a time. In others, it may be twenty or thirty. Are the project teams making progress? Is the coaching they're receiving beneficial to them in terms of helping identify the most critical uncertainties to pursue immediately? Are the business teams receiving signals of market enthusiasm? All of these metrics are appropriate for incubation.

Conclusion

Incubation is foreign to large, established companies. It's what start-ups do very well. Networking, learning, and redirecting with agility are all characterizations of start-up companies that have breakthrough offerings. Companies simply have not built incubation into their innovation functions. But they are beginning to do so now.

Next, we examine the last of the three building blocks: acceleration. Once a business appears to be not only viable but with a high potential, it's time to invest in its growth—explosive growth, in fact. Acceleration is far different from discovery or incubation.

Questions for You

1. What are the challenges of effective incubation in your firm? Specifically, what challenges are you experiencing with incubation processes? Career paths and talent development? Resources? Governance? Metrics?

2. How do these challenges differ for aligned and unaligned breakthrough opportunities? Which are easier to incubate? Why?

3. What incentives exist for the organizations currently responsible for incubating young breakthrough businesses?

4. What are you most pleased about with regard to the incubation competency in your firm? What are some incubation practices your firm has put in place that are most helpful?

5. What modifications could improve incubation of breakthrough innovations in your firm?

5

ACCELERATION

Gathering Steam and Building Critical Mass

Incubated projects need to make progress if they are going to get to the market successfully. But what does this mean? If you've been successful with discovery and incubation activities, you've developed a pipeline of projects, and several of your portfolio businesses appear to be highly promising. They've got a number of interested customer partners, across several different applications, and they're generating early revenues. It's clear that this is just the beginning, because additional opportunities may be emerging. Is it time to transfer them to whatever operating unit will ultimately run them?

No! Absolutely not, because you are missing the third building block of the DNA model: acceleration, that is, building the business's critical mass. Acceleration is a commitment to those businesses that, based on the experimentation done in incubation, truly appear to be breakthroughs. Every business that's being accelerated is being invested in heavily, with the expectation that it will

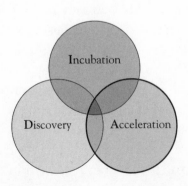

be one of the next major growth platforms for the company. Why is acceleration necessary? Because of expectations. Mismatched expectations. Let us reprise a story from our first book to demonstrate the problem.

IBM's Challenges with Developing the Silicon Germanium Business

In the late 1990s, Bernie Meyerson, a research fellow at IBM, knew he had a tiger by the tail. He'd developed a new chip that could process analog (rather than digital) information. It could do this because the material it was made of was silicon germanium rather than silicon alone. The additive properties of germanium enabled new materials characteristics and opened many new possibilities for IBM, including the telecommunications market. Meyerson and his team worked on the device over the next five to six years and developed a number of market applications with interested customer partners.

In time, the project was transferred to the microelectronics business unit for commercialization. The program manager in microelectronics assigned to head up the project, Mike Concannon, was discouraged. There were as yet no sales, no customers, and no reliable sales forecast. The opportunity that drove Meyerson and his team seemed like pie in the sky to Concannon. "I can't believe they sent it over to me this early," he told us. He directed Paul Cunningham, the new business development (NBD) person on the team, who'd transferred over to microelectronics division from R&D with the project, to develop a business plan and sales forecast for an upcoming annual budget review. Cunningham resisted because he knew that nothing he drew up would be realistic. He couldn't have confidence because too much uncertainty remained regarding both the market and the technology for any given application. But Concannon insisted, and so Cunningham complied. The business did not meet quarterly projections, and Cunningham was "sent to the penalty box," as he told us. He eventually resigned from the company. Concannon drew up a new list of potential applications, none of which were the ones Meyerson and Cunningham had identified or pursued, and started again. Why? He was skeptical of the conclusions that the R&D team had drawn regarding the various application probes that had been undertaken.

Meyerson said, "I kept dreaming . . . everybody kept dreaming . . . about the home run plan. . . . The focus was on 'how do you hit a home run?' because people kept saying, 'Oh cripes. After four years it's only going to be $250 million. You need to make it a billion dollars!' And so, we'd look at more applications. . . . I knew in my blood that the biggest application would take ten years. We knew that in the cell phone industry, the value proposition was clear, but companies that are billion-dollar businesses are not going to risk their business on new technologies . . . so we went to smaller applications that could prove our value."

For the first couple of years that the business was in microelectronics, its budget was meager—too meager to do what needed to be done and too meager to follow up on the applications that were being developed in R&D when it moved over. In fact, we learned that the silicon germanium chip business was nearly dropped from the Microelectronic Division's product portfolio plan because the projected sales for the first applications did not meet the required target for a new product launch. Bernie had to intervene from within R&D to save the project's budget and modify its metrics to reflect reality. Today the silicon germanium platform is a major business within IBM's Microelectronics division, an industry standard, and a critical part of the revenue stream. Yet a misunderstanding of how markets develop nearly caused its demise.

Meyerson's story is not unique. In fact, fully half of the breakthrough innovation project teams we studied that have been moved into operating units were either returned to R&D after it became clear that much work remained to be done, or their program managers were either fired or removed from the project because they could not meet expectations for sales growth. Once projects reach commercialization in operating units, they have to produce. Operating units are evaluated on short-term time frames, and their people are rewarded based on recent sales growth and profitability. Most business proposals that emerge from incubation have a long way to go before they can fulfill those criteria. It's not that there's anything wrong with them. It's just a fact.

Why Companies Need Acceleration Capabilities

Operating units succeed under conditions that nascent breakthrough businesses cannot deliver on. But all too often, early breakthrough businesses are expected to blossom in the operating unit's environment. Under most circumstances, this just won't happen. So what are the mismatches? They're the differences between the fledgling business's status once it's been incubated and the status of a business that a division needs to receive, as we portray in Table 5.1.

When companies move a fledgling business into an operating unit too soon, it withers. All that money invested in discovering and developing the business concept goes down the drain because receiving units cannot invest in developing the business further. They may not understand it, or they may not have the patience to invest. Their performance is measured on profitability and sales growth, not on investing to develop young businesses. There is still too much uncertainty with the fledgling businesses to warrant an operating unit's attention, the way most companies measure operating units' performance. (This is not the case universally. Some very large companies expect operating divisions to invest in their own future growth. But by and large, even companies that say this do not measure their operating units' performance this way.)

The aim and design of an acceleration function is to ramp up fledgling businesses to a point where they can stand on their own relative to mature business platforms and to operating management's performance requirements in their ultimate homes, whether existing business units or wholly new ones. It's all about critical

Table 5.1 Differences Between Acceleration Businesses and Needs of Receiving Units

Characteristics of Fledgling Breakthrough Businesses	Needs of Receiving Units
Several application market paths planned and tested	A clear set of market segments with chosen targets
Several customer partners	A qualified list of customers
First revenues from initial customers	A predictable sales forecast
Supplier partners identified and initial work begun	Suppliers and other partnership agreements in place, with well-defined performance expectations
A manufacturing and operations model identified and designed	Predictability regarding manufacturing yields or service delivery output per man-hour invested
A business model that fits the needs of the initial application markets being pursued, regardless of its fit with the company's current systems	A business model that leverages current assets and relationships
An analysis indicating that at certain volume levels, profitability will most likely occur, all else being equal	A well-understood cost structure and a known and favorable cost-volume-profit relationship
Hypotheses regarding a path forward for follow-on applications (insofar as this involves market experimentation)	A continuous improvement plan, feedback system, and plan for the business's follow-on products and markets
A passionate founding team with added new business creation specialists	Clearly defined roles and responsibilities, staffed with personnel who understand ongoing operations

mass of customers, business opportunities, operating assets, and people. Whereas incubation reduces market and technical uncertainty through experimentation and learning, acceleration focuses on building a business to some level of predictable sales. It involves developing the necessary infrastructure for the business, including a management team, marketing capabilities, manufacturing or operations and delivery systems, and the associated network of partners. It involves executing on the initial entry application but also planning and developing follow-on applications and markets. The manufacturing or operational model should be developed in acceleration to the point where yields become acceptable and a pathway to profitability is demonstrable. According to our participating companies that have invested in acceleration capabilities, it's about escalation rather than experimentation (which is the basis for incubation) or exploration (the basis for discovery). It's a move from a focus on pursuing early customer leads to a focus on developing predictable sales forecasts and a growth plan for exploding top-line revenue. Finally, the acceleration function introduces its potential businesses to the harsh realities of the world of ongoing operations. Traditional purchasing processes, traditional metrics, and traditional bargaining for resources are introduced though not yet imposed.

Acceleration is not free of risk and uncertainty. It's not a straight shot to a profitable business. Consider Airbus's A-380, the world's first superjumbo airliner, capable of more than eight hundred passengers. It was truly a breakthrough but fraught with difficulties. The Airbus team had done many of the right things for incubation. They'd worked out innumerable technical challenges, developed a cadre of partners to codesign and comanufacture the aircraft, and conducted a successful demonstration flight and a follow-on series of test flights over two years. And they had an impressive backlog of orders. But when it came time to ramp up and begin delivering on those orders, new problems surfaced. One known technical problem with the wiring was more difficult to fix than anticipated. In addition, operational issues surfaced. A-380 production is spread across four countries, causing logistical problems. (For example, wiring is done in France but then the aircraft is flown to Germany for wiring completion.) Couple these issues with the emergence of currency disadvantage (the euro has strengthened dramatically against other currencies since 2000), the emergence of massive financial shortfalls (being more than two years late has led to cancellation of many passenger aircraft orders and consequent loss of profits, the complete cancellation of the freighter version, and the potentially fatal delay of the smaller A-350), and governmental interference in employment alloca- tion (all four sponsoring governments insist on specified employee numbers, leading to excess workers), and one can see that accelera- tion is still fraught with uncertainty—certainly beyond what most operating divisions could absorb. Experimenting and implement- ing are clearly two different things, and neither of them fulfills the needs of an operating unit to help it run smoothly.

Why is an acceleration function necessary? Business opportu- nities that are forced to focus on profitability too early may under- leverage the platform of opportunities before them. They may revert to a product development project orientation and relin- quish the possibility of creating a new line of business. In addi- tion, breakthrough innovations need to develop business models that are appropriate to the business opportunity, which may not necessarily align with the designated receiving unit's operating model. The new business's model must have time to take shape and become institutionalized to the point where it will not be forced to conform to the dominant model. How do we know this? We watched several companies, in fact many companies, struggle with these issues. Several instituted accelerators. Kodak's was the best developed we've seen.

Kodak's New Business Accelerator

The mission of the Systems Concept Center (SCC) at Kodak's innovation hub, founded in 1994, was to develop systems-level concepts that cut across current business units and could become whole new platforms for growth. The founders, Gary Einhaus and José Mir, came from the technology organization and were cre- ative, prolific inventors and highly revered, even loved, by those who had worked with them. They built an organization that grew to seventy people across three locations, but was primarily centered at the company's worldwide headquarters, near its central R&D Group, in Rochester, New York. The SCC was populated with highly creative individuals, most of whom had strong technical backgrounds. Several who'd built businesses in the company were eventually brought in to help shepherd the ideas through their commercial paces. Einhaus developed a three-phased process for developing and nurturing a portfolio of ideas into business concepts. The alpha function was about creativity and generation of ideas and discoveries. Brainstorming, legwork, and interaction with consultants and R&D researchers fed that idea generation activity, and many of the SCC employees were highly engaged in alpha activities.

As an idea took shape and the SCC venture board scrutinized it for commercial possibility, a beta team would be assembled. Beta was about developing the project idea into a business concept, much as we described in Chapter Four. Those ideas, remember, were expected to be stretch ideas, thereby not really fitting in any of Kodak's five operating unit markets, production systems, or delivery systems. When concepts had matured enough in beta and demonstrated potential in the market, they were moved into one of the operating units that was the closest fit. But it became apparent over time that very few of these ideas were getting the investment they needed once they made the transition to the operating unit. The only ones that were making it to the commercial market were those that served the same market through the same distribution channels as the operating unit's model. But that wasn't the SCC's mandate. Its mandate was to develop new unaligned and multialigned opportunities.

At that time, Einhaus moved on to become the general manager and vice president of the new digital businesses in Kodak's Professional Photography Business and Nancy Sousa was appointed the new SCC director. The transition problem was a continuing plague. Sousa told us, "I need a landing zone for the projects that the business unit does not feel comfortable with. If I transfer these projects too early, the business unit leadership lets them die.

I need a place to grow them until they can compete with ongoing businesses in the current operating units for resources and attention." She ultimately convinced her boss, chief technology officer Jim Stoeffel and his boss, chief executive officer Dan Carp that this was an important addition to Kodak's innovation system.

The accelerator was to formally report to Stoeffel, but in fact was run by a broader-based governance board of the top five corporate officers: Carp, Stoeffel, chief marketing officer Carl Gustin, chief strategy Officer Mike Martino, and Marty Coyne, corporate executive vice president and group executive of Kodak's largest operation, the Photography Group. This very senior team (labeled the NEXT team) determined the composition of the accelerator's portfolio, held quarterly progress reviews with each young business, and decided when any one of its portfolio businesses was ready to move into an operating unit or to become a stand-alone business unit if that seemed appropriate. The entire operation was funded from corporate funds to prevent any instability in funding or bias in decision making that could creep in if business units were involved in the funding.

The accelerator was designed to accept young businesses from the SCC as well as from other places, including new business development (NBD) groups in the divisions that had identified opportunities that didn't fit with the division or were too early for their investment, as well as from the Kodak Ventures group, which was making investments in small start-up organizations. The goal was to have a portfolio of five to eight young businesses in the accelerator at any point in time.

After a prolonged search, Kodak hired Larry Henderson as the accelerator director. He had been involved in new business creation within large businesses (Motorola and Lucent) as well as a number of entrepreneurial start-ups over the course of his career and joined Kodak with a wealth of experience and lots of enthusiasm. He established an infrastructure including a leadership team, a set of processes, and a governance model for the accelerator over the next six to twelve months. He developed internal networks with purchasing, legal, human resources, and many other staff functions so he could educate them about new business creation and leverage their resources as needed. In fact, he had a stated goal of leveraging other staff functions at a ratio of one thousand to one (get one thousand people in other functions in Kodak to help with acceleration for each staff person he had). Henderson helped cultivate projects within the SCC and other sources to feed his pipeline. In conjunction with his two leadership team members, he worked with those businesses to help define and clarify their strategic value to the market and to Kodak. Finally, he managed expectations of the NEXT team regarding how each of the businesses in the accelerator could affect Kodak's direction and future health.

The accelerator leadership team worked with each business to hire a general manager, whom they referred to as the CEO, with industry expertise and high growth experience, and identify other relevant roles needed for the young business, including chief technology officer, chief operations officer, and chief financial officer. They added business and marketing capabilities to the team as well, since most of the young businesses had been initiated by people with a technical background but less business experience. Finally, the accelerator leadership team coached each of the young businesses regularly.

Each young business was expected to develop an advisory board. This group was not a board of directors and did not have decision-making power over the business, since the accelerator leadership team and senior-level NEXT team were in place for those purposes. Advisory boards had three to five members and included experts from outside Kodak to help the young businesses gain entrée into the new industries and professional groups that they were targeting. This external representation on the advisory boards was viewed as highly important, given that the new businesses were supposed to be just that—new to Kodak—and therefore could not rely on Kodak's traditional networks of customers, distributors, suppliers, and development partners. Each had to break new ground, and one role of the advisory board was to facilitate that. In addition, the advisory board could help with internal political issues and operational problems, such as how to gain access to necessary manufacturing resources that were controlled by divisions.

It's well known that Kodak has suffered years of declining revenues and threats to its core businesses and was slow in responding to the digital revolution. It took major change at senior levels of the company, including replacement of the CTO and CEO in the first half of 2005. Both of these new leaders had years of experience at previous positions at Hewlett Packard, and their belief was that new businesses should be accelerated within the business units. So the end of this story (for now) is that the accelerator, after it had developed its systems, its people, its process, its governance model, and its pipeline, was shuttered. At the time, it had a portfolio of ten businesses (larger than it wanted). Nine were moved to business units, and within six months, all had been discontinued. The remaining one was hidden in R&D because it was so highly unaligned with any current business unit that there was no logical home for it. It remained alive and struggling as of the last we know. Gary Einhaus, now back from his stint in the Kodak Professional business unit and serving as a research director in central R&D, reflected about the impact of the accelerator, "The accelerator's existence allowed us to consider more unaligned and stretch opportunities."

While the end of the story for Kodak may seem sad, it isn't. Kodak had to undertake a dramatic transformation in its entirety, and so its capacity for breakthrough innovation beyond that transformation to the digital world became highly constrained. We all hope it'll open up again as Kodak inches its way back.[1] In fact, much of the innovation capability embodied in the Systems Concept Center and even the accelerator was incorporated into the newly chartered Kodak Research labs focused on breakthrough innovation under Gary's leadership. Gary continuously refers to the "innovation principles," and recognizes that they can be imported to different places and, indeed, different levels of leadership in the company.

Now, Kodak has experience in building and running an acceleration capability. As long as they don't forget it, they may be able to resurrect it if needed. Once Gary Einhaus, José Mir, and Larry Henderson and their teams leave or retire, it's lost. That's why an innovation system needs to be institutionalized as a function. You can fire it up and damp it down, but if you turn it off, all of the learning will be lost.

The Acceleration Management System

If you want to develop acceleration as a building block of your breakthrough innovation capability (BIC), you must consider all of the elements of the management system, which should be familiar by now (Figure 5.1). Let's examine them one by one and consider how they play out for acceleration.

Figure 5.1 Management System Elements

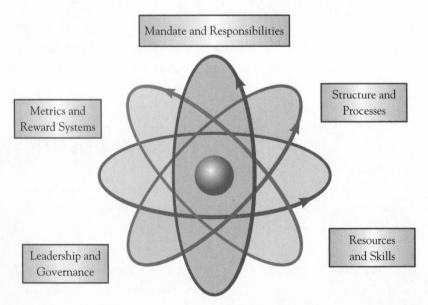

Mandate and Responsibilities

Structure and Processes

Metrics and Reward Systems

Resources and Skills

Leadership and Governance

Mandate and Responsibilities

The mandate of the accelerator is to jump-start growth. Intel's new business initiatives group was tasked with developing unaligned and multialigned opportunities. At Sealed Air, acceleration was defined as developing fledgling businesses that ultimately could fit into a business unit (though not perfectly; there could be a force fit) but needed continued investment. The function's particular mandate, as we have seen with discovery and incubation as well, drives lots of decisions. For in the case of acceleration, it will have a major impact on the accelerator's location in the organization.

The acceleration team requires three sets of activities: managing the relationship of the accelerating businesses with the rest of the organization, helping the nascent new breakthrough businesses gain critical mass, and joining the rest of the BI leadership team to educate the mainstream organization about the role and contribution of the this function.

Manage the Relationship of the Accelerating Businesses with the Mainstream Organization. The mother ship is steaming along. It may not be preparing to absorb a new division, or one of its divisions may not be preparing to be stretched to incorporate a new business that may appear like a stretch fit or even a misfit. In addition, businesses that are scaling need the attention of the staff and operations groups that provide resources to the rest of the organization. When young businesses experiment in incubation, there may be pilot facilities or dedicated staff associated with the innovation function to help with hiring, purchasing, legal agreements, and all the other activities necessary to accomplish any task. But when businesses are scaling, they cannot rely on favors or special services. They need the attention and resources of the rest of the organization. So the accelerator team must work with the mother ship to accomplish two objectives on behalf of its new businesses: garnering resources for scaling and preparing the mother ship to absorb the new business:

• Garnering resources for scaling. Young businesses need production facilities, purchasing support, human resource, and legal support. The acceleration function should not be an exception to these groups' activities. Larry Henderson at Kodak lobbied and won the right to have a dedicated person on Kodak's corporate legal staff, in human resources, and in the purchasing function available to help the new businesses with their needs. At Sealed Air's Cryovac division, manufacturing planning meetings include a representative from new business development, which incubates and accelerates growing businesses. That way, new businesses are granted regular time on the manufacturing line rather than being put in the position of requesting time as a favor. Similarly, Ohmid Moghadam, manager of the strategic programs group in Intel's central R&D technology management group, found individuals in the legal department who would commit to helping meet his group's needs for partnership arrangements, licensing deals, and other legal issues as they examined, invested in, and nurtured novel business opportunities. Ohmid was able to convince the director of the legal staff to add these activities, foreign as they were, to their performance metrics to help institutionalize the requirements to help with the company's innovation agenda along with its operational excellence agenda.

• Helping the mother ship prepare to absorb the graduating businesses. This requires continuously checking that the planned receiving unit is still the appropriate one, since the fledgling business's opportunities will surely evolve as they engage more and more with the market and their strategic fit may migrate from that of the company. The role of the accelerator leadership is to constantly monitor that fit and work with senior management to ensure the fit is still healthy or, optionally, redirect the young business's strategy in the light of its continued learning and experience.[2]

Accelerator leadership plays an important role in helping receiving units (or corporate leadership, if a new division is to be formed) work the new business into their planning and budgeting process and considering with them the timing of transitions and the impact on the pacing of the businesses within the acceleration portfolio. At MeadWestvaco, as with many of our other companies, some business units were more interested in growth through new businesses than were others. This difference in receptivity among business units helped prioritize resources and attention among the breakthrough innovation team staff members onto some businesses and back burner others within the acceleration group. At Intel, Angela Biever, director of new business initiatives, spent a large proportion of her time meeting with business unit leaders to time the graduation of her young businesses to match the budget and planning cycles of those landing zones that had to take over responsibility for their ongoing operations.

Help Nascent Businesses Gain Critical Mass. Young businesses in acceleration are investing to build the business and its necessary infrastructure. By investing we mean building an organization, an ecosystem of relevant partners, a manufacturing and delivery infrastructure, and a growth platform of many new leads. Focusing and responding to market inquiries and following them up are critical. In addition, beginning to design and institute repeatable processes such as order fulfillment, delivery, and installation, as well as customer service and support, becomes an important aspect of acceleration so that the pathway to a profitable business model can be demonstrated.

The accelerator leadership team can help its businesses in gaining critical mass through coaching on strategic clarity, development of the young businesses management team, and brokering relationships for them within and outside the company as necessary.

Educate the Mainstream Organization About the Role and Contribution of Breakthrough Innovation. The amount of time, effort, and skill required to build positive regard for the innovation function throughout the company is mind-boggling. We discuss this in detail in Chapter Seven when we describe the role of the orchestrator, but there is a management role, as we noted in Chapter Four, for each of the building block functions to play in supporting the orchestrator's effort to develop credibility for the innovation function.

In the case of acceleration, the acceleration leader can track and note every form of success for each of the accelerating businesses. Even fledgling businesses that never generate blockbuster sales may (1) stimulate whole new platforms of business due to new market or technological knowledge and networks they cause the firm to develop; (2) generate knowledge or new processes that are adopted by ongoing businesses, thereby improving the profit picture for many current product lines (our companies refer to this as "uplift"); or (3) benefit other portfolio businesses that are being incubated or accelerated at the time due to learning that impacts across the portfolio (our companies refer to this as "spillover").

An example of the first type is DuPont's Biomax®, the first in a family of biodegradable polyesters that DuPont has since commercialized. Biomax® was one of the projects that we studied from 1995 to 2000. DuPont technical leadership identified it to us as a potential breakthrough innovation. The Biomax® story is lengthy, but suffice it to say that after fifteen years of ongoing development and commercialization attempts, it has been a disappointment.[3] It has not brought blockbuster-level sales to DuPont.

However, the Biomax® experience caused DuPont to recognize the biodegradable opportunity space, engage with customer-partners across a variety of applications, experiment with many types of materials and manufacturing processes that would work under certain conditions, and launch, over time, a family of products around which the company has formed an entire business unit. Using traditional tracking techniques, one may not recognize the "success" story that Biomax® really is for DuPont. That is why the accelerator team has to constantly consider and track all the ways that success manifests itself in the breakthrough innovation domain.

Structures and Processes

Structures for Acceleration. We've seen companies experiment with many kinds of structures to handle acceleration. Since formal structures for handling acceleration have been in the ascendant for management only in the past few years, experimentation is understandable. Here's what we've seen from among the twelve companies we've been studying most closely. The numbers add up to more than twelve because companies have changed their approach over the four years that we've been studying them.

Two companies created specific accelerator groups. One of those reported to a governance board of corporate officers, and the other resided in a business unit but handled the entire portfolio of accelerating businesses, aligned or not with the unit in which the accelerator resides. This occurred because only one individual emerged among the identified set of people who appeared interested in and able to lead a portfolio of emerging businesses. Rather than have portfolio businesses make the transition to accelerators with incompetent leaders, the incubator leader elected instead to move them all to one competent accelerator leader, whether or not the fit was right organizationally. The incubator leader pledged support to help this person with the misfit businesses, but the time and attention the accelerator leader gave to the misfit businesses eventually was reallocated to those that fit the business unit since his performance measurement was done by the unit leadership. (See how mismatches among the elements of the management system can mess things up?)

Three companies have had a single division in the company responsible for developing fledgling businesses that are unaligned with current operating units, from start to finish (the entire DNA system was located in one division). Eight companies expected fledgling businesses to accelerate in the operating units without the protection of an acceleration function.

Four companies did not use any organizational structure but instead relied on senior leadership oversight to ensure the businesses did not suffer from lack of attention or resources. In some cases, business units are expected to share funding on the project as it gets ready to make its transition. Four other companies moved the fledgling businesses to an operating unit but did not benefit from senior leadership's oversight. In only one of these cases did a new business actually ramp up, and it was pushed by a well-known and well-regarded project champion who left the BI group to run the new business. In addition, three of our twelve companies have developed unaligned new businesses to the fast-revenue growth stage (through acceleration) within R&D on occasion.

Clearly acceleration is being strong-armed, handled on an exception basis, or ignored in most organizations today. There is no system. Companies have not arrived at an infrastructure to handle growth and scale-up of businesses that do not fit the conventions of the operational excellence system. Most companies try to accelerate breakthrough businesses within business units and tell us that they'll be patient. That is why so many breakthrough projects that finally do move to the mainstream operating units never make it. They're underleveraged and underresourced. They've come so far, and yet the company unwittingly scuttles the investment. By instituting an acceleration function, breakthrough businesses can be spared from this fate.

So what's the right answer? Where should the acceleration function be located? Table 5.2 lists the advantages and disadvantages of the options available to companies. So long as companies are aware of these, they can align their acceleration management system.

In one of our participating companies, the BI orchestrator was toying with the following well-thought-out idea for ensuring the breakthrough innovation businesses would be accelerated within business units and decided that it had merit:

- When a project was initiated in discovery, the orchestrator considered appropriate eventual homes for the ultimate business that could have potentially emerged.

Table 5.2 Acceleration Location Options and Trade-Offs

Alignment of Breakthrough Innovation Business with Existing Businesses	In Business Unit or Divisions	At Corporate Level
Aligned	*Pros* Workable if acceleration-related metrics are imposed on the division by corporate and if all business units are expected to accelerate businesses. Eases acceptance of the business if the business unit is required to invest in it earlier. *Cons* Metrics in business units are mismatched with those appropriate for acceleration. Big successes for BIs in early acceleration are viewed as rounding errors in the division's current planning horizon. Too tight a link with a business unit causes pressure toward incrementalization.	*Pros* Funding with corporate money protects business units by not hindering their profit picture for far-future businesses. Protects against lack of attention due to urgent matters associated with current customers and product lines that may occur within business units. *Cons* Problem of not-grown-here turf wars. Transition to operating division can be difficult in terms of budget and planning cycle, as well as availability of appropriate personnel and facilities. Can't attract the business unit's interest. Extra coordination with business unit required to avoid or manage duplicate visits to customers and channel members from those representing the accelerating businesses and traditional sale force representing the business unit's product line.

(Continued)

Table 5.2 *(Continued)*

Alignment of Breakthrough Innovation Business with Existing Businesses	In Business Unit or Divisions	At Corporate Level
Unaligned	*Pros* None *Cons* Business units not designed to invest in growing businesses that do not fit their strategic intent or operating model. Force-fitted businesses will constantly battle for resources that the business unit is unwilling to invest in if the business unit cannot find other ways to leverage those resources (for example, new manufacturing equipment to accommodate a different product process from one the business unit traditionally uses).	*Pros* Allows freedom to try out business models that may not fit any current operating unit. Allows focus on a portfolio of accelerating businesses, with decisions on when and how to pace them to be made based on big picture. Forces continued discussion with senior leadership regarding link with strategic intent. *Cons* Requires investment in a new infrastructure—an accelerator at the corporate level.

- He informed those business unit leaders at the highest levels about the project but did not allow them any oversight, for fear they'd incrementalize or force-fit the business into their current systems.

- The project work was carried out and incubated within his innovation group, funded by corporate money. Divisional manufacturing resources were used for trial runs but were reimbursed by corporate funds.

- Once the incubated business was achieving sizable revenues (in the multiple millions of dollars) and established, or regular, and growth appeared promising, the business unit was expected to take the business over and reimburse the corporate innovation group's expenses. It was treated as if the business unit were acquiring a start-up firm. If the business unit stalled and expected the corporate innovation group to continue to reduce uncertainties and the future risk around the business, the price rose. The orchestrator told us, "Our company is better with acquisitions than with innovation. If we treat this like an acquisition, the business unit will know what to do." He figured that the risk of the fledgling business had been lowered enough that he could expect the business unit to accelerate it. In addition, since the business unit was required to pay for it, it had already invested enough that it would not neglect to leverage the opportunity into the full business potential that it promised. Lowering the risk for the business unit worked. Charging the business unit worked as well. Neat idea.

Acceleration Processes. In every project we've studied, after numerous application probes were analyzed, a small number were targeted for development. This required a focused strategy and a shift from probing to developing a niche-entry application for market creation purposes and a plan for follow-on applications. Markets broaden and new markets emerge as they learn about the technology, sometimes faster than the business team can respond.

The processes for acceleration are quite different from those of incubation. Incubation processes experiment, test assumptions, and create options. Processes for acceleration are about focusing on execution and responding to inquiries as the market learns about the breakthrough's many potential applications. The temptation at this point is to follow every lead and continue to generate options. But that's a problem. It's easy to get spread too thin and get too diffuse. The young business's leadership, in conjunction with its advisory team, must devise a plan of cascading markets and execute it. All the time that this is going on, the leadership team for the young business is building the organization and infrastructure. Planning for growth is a big part of this process. Selecting applications and customers are important decisions at this time. The business's leadership team must decide which opportunities *not* to accept.

Transistors were first used not in radios but in hearing aids and missile guidance systems. Lasers were first used in measurement, navigation, and chemical research.[4] Over time their market broadened to include music recording, surgery, printing, fabric cutting, and communications. DuPont's Surlyn is another example of market evolution for a truly breakthrough innovation. It took twelve years for Surlyn to break even, starting as a coating for golf balls and a replacement for high-heel tips, and eventually growing into a multimillion-dollar business with a myriad of applications, including food packaging.[5]

The same evolution can be seen in other cases we've studied. At GE, the digital X-ray team began to focus on a number of new initiatives once the market came to understand the technology's potential. Research scientists from other firms soon began contacting GE with application ideas. At Analog Devices, the project moved from providing chips to the automotive market at low prices and high volume but no profit to selling chips profitably for computer games. It then shifted to box games like Nintendo and Sony PlayStation, where volume and margins are even higher, and then more recently to the Nintendo Wii, where it has finally achieved the volumes envisioned for a major business. Subsequently Analog Devices began receiving inquiries about using its sensors in sporting goods and other applications where vibrational changes need to be noted. Texas Instruments' digital micro-mirror device business struggled for years after its launch in 2000. As the market grew familiar with the breakthrough, however, it has fundamentally changed the projection industry. TI eventually had to set up an office for evaluating the flood of inquiries about new applications for this technology.

As managers begin the task of selecting the markets to pursue, they're fraught with conflict. On one hand, they feel pressured to produce game changers, especially after millions of dollars have been invested. On the other hand, markets are not yet educated, and technical bugs may remain. One challenge associated with this pressure for big wins is that acceleration teams can diffuse their focus. In several cases, we witnessed the expansion of numbers of application probes to an unmanageable level once the project moved to a business unit, as this program manager explained:

> One of our problems getting started was in our effort to make money quickly, the team was too diffuse. They wanted to do everything. Wired applications . . . sonic applications, they were focusing on storage . . . and each one of those applications requires different knowledge and a different culture basically, so when I came in, probably six or eight months into it, we redefined it. I told them that . . . the only business we were going to focus on was wireless. So I focused the team and focused the energy on customers, and applications, and knowledge rather than having it so confused.

The program manager of a different project echoed the same warning:

> It was apparent to me we were trying to do too many things. And so consequently we weren't moving any of them forward as fast as anybody really wanted. . . . We're going to have to choose the areas

that we think we're going to be successful in earliest, and we must choose the right number of customers to work with. So we scaled down from forty initiatives to four market segments and a total of eleven customers.

We and others note that new businesses do not get built quickly. Sometimes the expectations of the potential magnitude of the BI business are so heightened that early market entries appear disappointing. The theme of interacting with the market so that the market learns about the technology is key in the acceleration phase. None of the projects in our first study achieved a killer application early in their commercialization phase.

Figure 5.2 shows the case of Analog Devices' accelerometer. The company's vision of the killer application was the automotive market, first for airbag detonation and then in other applications.

Figure 5.2 Application Migration for Analog Devices' Accelerometer

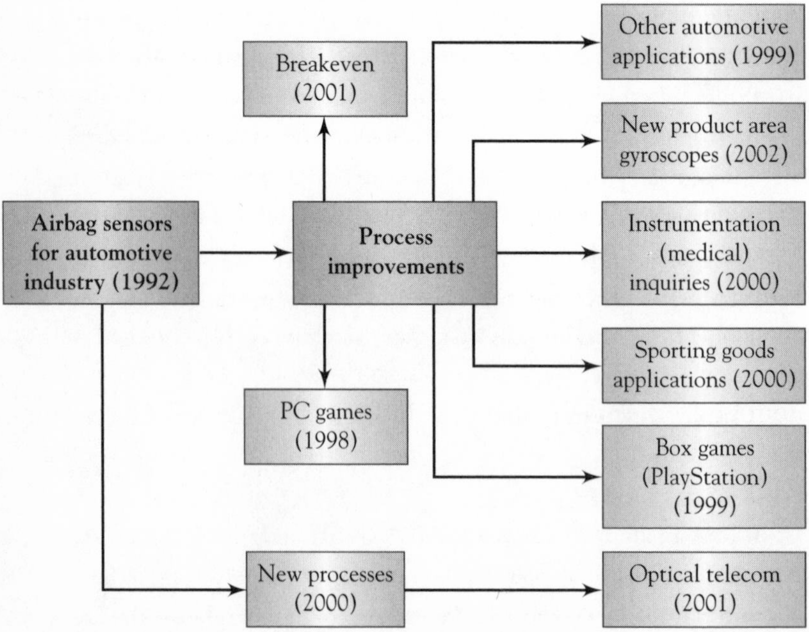

The company sacrificed early profits to gain those volumes and manufacturing learning curve advantages, but the market actually evolved way beyond that and brought profitable applications to Analog because they saw the potential as Analog's accelerometer technology became understood. The point of Figure 5.2 is that application migration occurs, meaning that an early entry application may be in a niche market, but others arise and seek out the innovating company to learn more. So focusing on any single application space and expecting a killer application is not connected to reality. It is critical to be open to inquiries from fields far removed from those originally envisioned, but to have a plan about the pacing of investment in them.

Resources and Skills

Of the three building blocks, acceleration requires the most money. Investing to increase a business requires major investment in capital assets (exceptions may include financial and service innovations, in which the initial discovery research is the most expensive aspect), as well as growth in personnel from team-size groups to entire business infrastructures. Several of our companies acquired firms or large divisions of companies at this point in a fledgling business's growth. GE's acquisition of Amersham in October 2003 to accompany its biotechnology advanced technology program's discovery and incubation activity catapulted GE into the business of personalized medicine.[6] GE subsequently formed the Health Systems division to capitalize on the combined strength of the infrastructure and market connections purchased through Amersham, along with the breakthrough discoveries and market opportunities coming from the advanced technology program (ATP) activity. Clearly accelerators likely cannot manage a very large portfolio, since every one of their fledgling companies is on a fast growth track, soaking up lots of funds. Indeed, in medium-sized companies, they may be able to support only one or maybe two projects in acceleration due to the significant amount of resources required.

Acceleration requires specific kinds of people resources as well. The skills needed for acceleration are those required for managing high-growth businesses. Therefore, accelerator staff, who help choose the general manager and leadership team for each of the accelerating businesses, will need acumen regarding managing high-growth businesses. To do this requires addressing a number of questions: Which opportunities are the best ones for our businesses to pursue, and which should be saved for future exploration? How can we push the economic model of the portfolio businesses to be more acceptable? What connections must be made between our businesses and others inside the company? In addition, the accelerator team, in concert with the accelerator leader, must consider the health, diversity, and pacing of the portfolio and be able to justify the portfolio's composition to senior leaders at any point.

Finally, several of our participating companies noted that ensuring a critical mass of people to execute acceleration of a BI business was a perennial challenge. At a time when follow-up and focus on execution are critical, underresourcing the potential breakthrough business is a sure-fire road to failure. Process and product development engineering, field support personnel, and customer service personnel are all necessary. Manufacturing oversight is critical as well, since, as one person told us, "the manufacturing process is likely new and must be 'bird-dogged.'" Finally, time must be reserved for the strategic planning to plot the future course of the business, given that opportunities may be flooding the gates, or not.

Leadership and Governance

The accelerator leader must have experience in developing businesses. The leader and the team must be able to work with the portfolio businesses as they experience the pains of growth. The role relies heavily on the leader's political and communications skills to describe in a compelling way the needs of the accelerator portfolio to the mainstream organization. She or he must be able to represent them in the ongoing battle for resources.

The accelerator leader should be a person with status and prestige who enjoys the respect of peers in the organization. This is required to work with corporate staff (public relations, legal, human resources) to obtain appropriate resources, as well as to work with business unit leaders to engineer the smooth transition of breakthrough businesses. The leader's ability to influence must be high, given that the portfolio of businesses requires heavy capital investment and a serious commitment from the company. This is not experimentation anymore; it is escalation—escalation of resources and escalation of commitment. Sometimes accelerating businesses fail, and sometimes they're killed, but more typically, businesses in acceleration have been through enough trial to give the company confidence that they'll be reasonable investments, indeed, exciting investments. Finally, the accelerator leader must have a keen sense of people to build, along with his or her staff, effective management teams for each of the businesses.

In terms of governance, decisions must be made about which businesses come into the accelerator, how long they stay there, what their ultimate home will be, and when that move should occur. These are major decisions requiring agreement among a number of senior leaders. Just as Kodak's NEXT team and Corning's Growth and Strategy Council were composed of the top senior officers of the company, so too will any accelerator function want a governance team that operates at the highest levels of the organization.

If the accelerator is at the corporate level, making decisions about a portfolio of businesses that are force-fits into divisions or may require the formation of a new unit somewhere within the company, then a team like Kodak's is required. If there are accelerators within each division ready to receive and develop aligned opportunities, the governance team will comprise the incubation leader and senior leaders of the division. For example, in addition to the senior leaders on the team, Corning's Growth and Strategy Council includes the four vice presidents in the company who control most of the technical resources and can bring those to bear, make priority and resource trade-offs as needed, and oversee the execution of the decisions made by the Growth and Strategy Council.

At MeadWestvaco, the president of new ventures and radical innovation, Don Armagnac, purposefully required that each of his accelerating businesses have a board of overseers, comprising industry experts as well as several senior internal leaders. Armagnac himself sat on every one of the boards. His purpose in adding senior leaders to the boards was twofold. The first, of course, was to help align the strategy of the new business with that of the company. But the second was very interesting: he wanted to expose senior leaders to the real world of early, high-potential-growth businesses. He believed that the education and experiences they gained as one of a set of advisers to a high-growth opportunity would be useful in their roles as general managers of large, established businesses as well. It was his way of effecting culture change at MeadWestvaco—by starting at the top.

Metrics and Reward Systems

The temptation to measure the acceleration function on traditional profit-and-loss measures is very strong in most companies. Most have systems in place to gather and report the data that way; planning is done on the basis of those kinds of results, and priorities are set that way. But once again, watch out: there are ways to measure the health and contribution of an accelerator function, but they must match the mandate of the function. Next we look at some possible measures.

Growth in Sales and Inquiries of Portfolio Businesses. Businesses in acceleration should not be judged using profitability-based metrics because they still require investment. Revenue growth, number of new inquiries for development work, or number of new customers added can be used as key indicators of success. We saw all of these metrics implemented in companies that understand acceleration as a function. The projects are held to performance standards, but their metrics are based on market activity rather than financial activity. As one program manager told us, "When you're thinking of a start-up, you can't measure things by internal investment or profits; you've got to measure revenue growth, even if that revenue growth in the beginning is not set and you haven't accounted for expenses. If somebody's willing to pay you a quarter of a million dollars to write a design on your chip, that should be an indication that they're serious."

In nearly all the cases we studied that reached the point of market entry, the markets chosen at the start, when rapid ramp-up was less important than rapid, inexpensive learning, were not the kind of "killer-application" markets envisioned for the technology when it was initially funded (Table 5.3). The entry application of GE's breakthrough digital fluoroscopy technology was breast mammography because it used the smallest, least costly, and least complex machine. But a chest scanner had greater market promise, and a full body scanner was the ultimate goal. At TI, mass-market

Table 5.3 Envisioned Market Entry Applications Versus Actual Entry Applications

Project	Envisioned Killer Application	Actual Entry Application
GE Digital X-Ray	Chest scanner	Breast scanner
Otis's multidirectional elevator	Will enable mile-high buildings	Moves prisoners underground
Texas Instruments' display projection system	Mass market projection equipment	Large-screen theater projection systems
Air Products	Large fabrication plants	Hospital medical equipment
IBM's silicon germanium chip	Cellular telecommunications	Global positioning systems satellites
DuPont's biodegradable polymer	Disposable diapers	Various packaging applications

projection equipment was set aside in favor of large-screen movie projectors when manufacturing yields proved too low to support the low margins of a mass-market device. Instead, large theaters and custom signage were the initial entry points for TI. We observe that in most of these cases, a number of smaller-entry applications led to killer businesses.

The implication is that metrics for acceleration must accommodate the reality that successes in small application spaces are really part of a larger program of innovation in a domain. Instead we observed projects that were criticized for not bringing in large revenues through a single killer application. Multiple paths to the market, not one superhighway, is the more likely scenario. Acceleration should be measured on sales growth, either within a killer application or, more likely, across a variety of markets.

The issue of metrics is quite difficult and remains problematic from one company to another. Remember that you get what you measure for, and so decisions about metrics become an important issue. Very few companies trace revenue streams for new businesses to see how they have contributed to paying back their investment. Perhaps more businesses should. Perhaps they don't because of the following issue.

Reduced Traceability. Letina Connelly, a member of IBM's emerging business opportunities (EBO) staff noted to us that it becomes difficult to trace the impact of high-growth businesses on the company. "Over time, these EBOs blend into the organization. Over time, it becomes harder to carve out the original EBO, because it's part of what we are," she says. "We, IBM, are becoming them." That's a powerful statement about organizational renewal. When the young business is influencing other aspects of the company by infiltrating additional lines of business that way, it's become a true engine of corporate rejuvenation. IBM used to have an Internet division. That was how it started in e-business. It was a forcing function to get the company invested in the Internet as an emerging market space. It has become part of the fabric of every product and service IBM now offers, so there is no longer a revenue contribution from IBM's Internet division. Today, there is no Internet division, as Internet technology is embedded within everything IBM does. Yet it's clear that IBM's investment in the Internet has surely generated a lot of revenue for the corporation.

Spillover to Other Platforms. Another signal of BI success is that business opportunities may move into explosive growth mode, and new applications may emerge that become important to other business platforms or generate new ones. As businesses accelerate, they spawn new discoveries and new opportunities that feed the early pipeline. The more the business interacts with the market and the market becomes familiar with the technology and the opportunity, the more the market can identify new aspects of the opportunity that may be valuable. IBM's first EBO in life sciences graduated from horizon 3 (incubation) to horizon 2 (acceleration). At that time, a new need was identified by those customers of the life sciences business in information-based medicine— a wholly different business platform that became a new horizon 3 project. Another one of IBM's early EBOs was focused on pervasive computing, a technology and service platform designed to provide mobile people with any information they need anytime, anywhere. Work was proceeding well, and revenues were reaching the hundreds of millions of dollars. The EBO was horizon 2, moving into the mainstream operations of the organization. But out of that experience, a new opportunity was recognized in the area of sensors and actuators, and a new horizon 3–level EBO was formed.

Uplift. One of our companies has had the strategic intent to produce breakthrough innovations in a general platform centered on photonics. At an informal get-together, one of the lead engineers happened to mention in passing an aspect of the progress that had been made. A high-level manager, to whom the conversation was not directed, picked up on the conversation and asked a few questions of the engineer. His interest piqued, the manager asked if they could continue the conversation concerning certain aspects of the photonics work in more detail in a workplace setting.

The upshot was that the work in photonics enabled the construction of a key control device for a very large-scale machine that would be far more effective than any that was currently in use. The photonic-based device led to a substantial increase in the efficiency of the machine, which produced large financial returns for end-use customers, so much so that the competitive landscape was substantially disrupted. Our company experienced substantial revenue and profitability increases in the affected businesses in both the retromarkets and the original equipment markets. This type of uplift forms one of the principal metrics for this company: "When I see stuff coming out of BI efforts and making contributions to previously unrelated activities and businesses, I know that things are going well. We don't explicitly look for revenue flows or impending business, although that's even better. When I don't see contributions to other businesses and to our customers emanating from a BI effort, it's time to take a hard look."

A number of our companies have experienced significant uplift of this type in existing businesses different from the BI business being pursued. One leader of a BI evaluation group flatly states that "whenever we checked, the uplift from BI alone is more than enough to justify its ongoing existence." The uplift and spillover can also be taken as examples of early harvest of revenues emanating from BI efforts, a mechanism for the generation of additional options of potential directions burgeoning out of the innovation efforts, and a natural generator of portfolios within portfolios. These are the types of results induced by participation in BI creation that often get overlooked because of management's inattention to the extent of their contributions or because of the inherent difficulty of valuing them on an accounting basis.

Number of Businesses Moved to Operating Units. Here the number of new businesses that have made the transition to an operating unit is compared to a stated objective. Make sure your objectives are reasonable. Acceleration is expensive.

Impact of New Businesses on Strategic Intent.   The point of having a breakthrough innovation capability is that it's the engine of growth and renewal for the company. In other words, what are the impacts of the new businesses on the company's fulfillment of or reconsideration of strategic intent? Are the businesses that the innovation system is generating doing that? Or are they more of the same old, same old?

Perceived Value of Acceleration Group.   Although there will always be a tension between the world of innovation and the world of operations, ultimately the world of operations has to perceive the world of innovation as helping to shore up its future promise. Are you getting calls from operating units? Are they working with you to shape your portfolio in such a way that it will benefit them? Are they fighting over you as a resource? Do they perceive that the businesses they're receiving from the accelerator are mature enough? Do they perceive that the businesses in the accelerator that are not a fit for them are exciting opportunities for the company? All of these are good signs and should be incorporated into a measure of the acceleration role's value to the company.

While acceleration of new businesses will change the dynamics of markets, breakthrough businesses are heavy investments. Acceleration is key to have realistic expectations about what they'll deliver and when, and then measure on those objectives. We cannot emphasize enough the benefits that cannot be measured in dollars because of the way companies' accounting and information systems are structured. Learning, uplift, spillover, and impact on strategic intent are benefits that most companies do not track. It's important to consider them, and the acceleration leadership team should make sure they're doing so.

Conclusion

The focus of acceleration, as the third building block of organizational competence, is to get projects to a place where they gain acceptance by business units. They need to be strong enough and robust enough to compete against the portfolio of other offerings the business unit spends effort on. To do this requires growth and demonstrable customers, sales and revenues, and a level of predictability—predictability of sales forecasts, cost models, manufacturing processes, and inventory handling models. Gaining this momentum is the goal of acceleration.

There are numerous pitfalls accompanying the acceleration set of activities that can be avoided for the most part with a clear understanding of what acceleration is about and how it should be managed and positioned within the BI system. Acceleration activities are designed to overcome the natural resistance by business units to new products or services that might depress short-term sales and revenues, since that is how most business units are evaluated.

With this chapter, we have completed our discussion of the basic DNA building blocks. We next turn to how these elements are related to each other and how they work together. We will see that interface issues as well as balance of effort issues among the three components can help or hinder the breakthrough innovation system's effectiveness.

Questions for You

1. What are the challenges of effective acceleration in your firm?

2. How do they differ for aligned and unaligned breakthrough opportunities? Which are easier to accelerate?

3. What incentives exist for the organizations currently responsible for accelerating young breakthrough businesses?

4. What are you most pleased about with regard to the acceleration competency in your firm?

5. What modifications could improve acceleration of breakthrough innovations in your firm?

6

THE DNA INNOVATION SYSTEM

Most companies don't appreciate the systemic nature of getting breakthrough innovation (BI) accomplished. Up to this point, we've focused on the building blocks that are necessary for a breakthrough innovation capability. Discovery, incubation, and acceleration (DNA) are surely necessary if a company hopes to get successful breakthroughs on the basis of anything other than chance, but they are not sufficient in and of themselves. The real payoff comes when they are managed as a system. If they are not managed as a system, with appropriate interfaces among them, as shown in the diagram above, the breakthrough potential of the company's portfolio will not be realized. This chapter focuses on the parts of the DNA system and how they work together as an integrated whole. We discuss the interfaces among the elements and how they should be coordinated.

We also look at the management system elements, now familiar to you, and how they play out at the DNA system level. There are system-level activities, in addition to each of the building block activities, that must be attended to. How do we know? Once again, we look back to our data.

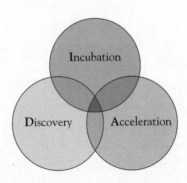

Our research team rated each of our initial twelve companies twice on its strength in discovery, incubation, and acceleration: once at the beginning of the study period and once towards the end. Firms were evaluated as having a high degree of competency in the area if they had developed some processes and invested resources into that competency's development and were having some success in that area. A medium level of competency was noted for firms that were either (1) having some success in the area despite a lack of formal recognition or investment in the activity or (2) investing in the activity in order to build competency and infrastructure. A low level of competency was noted for firms that were not engaged in the activity at all.

What we discovered was that few companies had high levels of competency in all three areas. This isn't really a surprise, since the concept of an innovation function is new and foreign to most organizations. But we were surprised by comments from those few companies that were rated at a high or medium level of executional excellence on all three building blocks: they were still frustrated by a lack of throughput in their innovation pipeline. When we listened to their challenges, we realized that gaps between discovery, innovation, and acceleration were accounting for these difficulties. This was particularly important for companies that didn't manage all three building blocks under the same organizational umbrella.

Only one of our companies rated high on all three building blocks, but its links between the discovery and incubation functions were weak. In follow-up interviews, our participants noted that they were having trouble finding new opportunities as the initial ones were maturing into credible businesses. They eventually hired an additional team member to work with R&D and other organizational units to help strengthen the flow of ideas into the BI management system.[1]

Shell Exploration and Petroleum (Shell E&P), another in the family of Royal Dutch Shell companies, had a thriving discovery group (familiarly called GameChangers) and other groups responsible for maturing the ideas coming out of GameChangers and then transferring them into the business units for acceleration. But no one was overseeing the transition, so potentially promising ideas were not being pursued. Some identified opportunities would get picked up by the incubator groups and others would not: it was their choice. Some ideas were breakthroughs, and some were incremental in nature. There were different pathways available in the company for different kinds of innovation (for example, a fast-track path to the business units for incremental innovations), but no one made sure these were being used, or used appropriately.

Ultimately the GameChanger group initiated an innovation assessment for the company and discovered these major loopholes in their system, so they brought all of the constituents together (the GameChangers, all incubator owners, and business units) to hammer out the details of which group was responsible for doing what. They created an oversight board (comprising the discovery, incubation, and acceleration leaders). The person in charge of managing the board kept a tracking system of all of the projects as they wound their way through the three phases, and the result is that projects no longer got lost. The governance team made strategic choices. Shell E&P's breakthrough innovation system became recognized within the Royal Dutch Shell family of companies and among innovation experts as one of the most effective innovation systems in existence today.

So it becomes clear that the DNA building blocks have to be managed as an interrelated and interactive system. This means that there are system-level management and coordination responsibilities associated with a breakthrough innovation capability in addition to these individual building block activities and responsibilities:

- Managing the links and interfaces among the DNA building blocks
- Managing the balance of resources and capability development across discovery, incubation, and acceleration

- Managing the health of the BI portfolio of projects
- Monitoring and adjusting the linkage and fit of the DNA system with the rest of the organization in accordance with its ever changing capacity for innovation

We'll take up the last issue in the next chapter, where we discuss orchestration. For now, let's examine these responsibilities that focus primarily on challenges that can arise within the DNA system.

Managing the Links and Interfaces Among the DNA Building Blocks

For the discovery, incubation, and acceleration functions to operate successfully, their interfaces have to be monitored and managed. This requires three major tasks: transitioning individual projects and businesses from discovery to innovation and then to acceleration; pacing the businesses appropriately through the innovation system; and ensuring that feedforward and feedback are occurring across discovery, incubation, and acceleration.

Project Transitions

In Chapter Five we described the difficulty IBM faced in the transition of the silicon germanium chip business platform from incubation to acceleration. The project lost momentum and organizational attention. That can happen when the group responsible for incubating and the one responsible for accelerating are different people. That's one reason we've noted the need for system-level oversight.

GE's chief technology officer, Scott Donnelly, worked directly with GE's business unit vice presidents to craft a strategy for new businesses that were spawned from the advanced technology programs (ATPs) initiated in GE's Global Research organization in 2000. This high-level oversight carried out by the heads of discovery and incubation/acceleration ensured that GE would leverage those platforms to maximize their business opportunities. At Air Products, Ron Pierantozzi's new business development group reported to the vice president of commercial development, but Pierantozzi placed one of his most experienced staff people in the R&D organization to ensure that opportunity articulation and elaboration (the business parts of discovery) would take place and feed the major innovation pipeline. Finally, Corning's high-level Technology Council and parallel group, the Growth and Strategy Council, oversaw projects as they moved through the innovation maturity cycle. The president, the chief technology officer (CTO), and the four vice presidents representing corporate research, development, engineering, and new business development, respectively, sat on both councils, so smooth transitions were ensured. At DuPont, business units begin sharing in funding of projects that are in incubation, so they become prepared to receive these projects as they mature. With business unit vice presidents sitting on the review board, these alignments are reinforced.

However, the reverse problem can also occur: transitions are not clarified because no distinctions are made among discovery, incubation, and acceleration. In fact, in most companies we've studied, discovery and incubation (less so acceleration) are overseen by the same people. In these cases, the problem wasn't a lack of linkages among these three so much as failing to distinguish among them. This means that teams that are doing discovery are left intact to engage in incubation, even though they don't have the right skills to do incubation well. The projects languish because team members don't know what to do or don't feel comfortable doing it, so they keep doing what they know best: technology development.

In one of our participating companies, the head of the discovery program ordered the discovery team not to engage with the market. The discovery team's job, she insisted, was to engage in scientific research. Yet the time came in their discovery work that they needed market information and needed to engage in incubation-related activities. They snuck out to visit the market against orders! They went to professional meetings and discussed the project with potential customers. They visited potential customers and talked with research and engineering staff who might understand the implications of the technology.

Some incubation teams don't move toward accelerating, because they love to explore and find exciting opportunities, but asking them to focus on that first one and follow it up to build predictable sales growth is like asking them to swallow a sleeping pill. One project team's new program manager, who was expected to jump-start the business's revenues, expressed frustration with how difficult it was to rein the team in: "It was apparent to me we were doing too many things. And so consequently we weren't moving any of them forward as fast as anybody really wanted. . . . We're going to have to choose the areas that we think we're going to be successful in earliest, and we must choose the right number of customers to work with. So we scaled down from forty initiatives to four market segments and eleven customers."

Another problem we see with system overseers who do not distinguish discovery from incubation from acceleration is their tendency to measure project performance with the same set of criteria. Yet it's clear that the criteria for measuring progress should change among these, as we've noted in the previous three chapters. So transitions of projects from discovery to incubation to acceleration must happen explicitly and must be managed. They cannot be ignored, and they have to be monitored once a transition has occurred to make sure the fledgling business is adjusting to its next set of challenges.

Pacing Issues

Discovery, incubation, and acceleration system managers may face pressure for an even, steady flow of projects through the pipeline. Our participating companies tell us that pressure to accelerate causes transitions from incubation to acceleration that are too early. For a project to be ready for acceleration, as mentioned in Chapter Five, it has to have a demonstrable path to profitability and a stream of first revenues. However, several companies have told us that this gating criterion is sometimes too difficult to meet; they make the transition of these projects anyway because the long and winding road is too long and the spotlight is on them. If management presupposes a certain time period associated with incubation, or discovery, or acceleration, then those expectations are set, and projects will be paced accordingly. This can be disastrous, since projects that are too immature when passed along to the next innovation building block will not be ready for the activities it demands.

Angela Biever, former director of Intel's new business initiatives group, told us that one of her biggest challenges was that her pipeline was not evenly paced. Intel believed it should be rather evenly paced, but she indicated that, at times, her portfolio profile is like a snake digesting a rat: they're all at about the same level of maturity. She incubates and graduates new businesses in chunks, she tells us, "because that's the best I can do in terms of paying attention to them. If they're all accelerating, I can help them all as a group. If they're all incubating, it's easier for me to help them. I have several graduating this spring, and then I'll start a fresh batch." At any one time, Angela has several projects in her pipeline, although she transitions or graduates a set of projects at a time to the next stage of development. The point is that she recognizes that each stage of development requires a different kind of management attention and guidance, which she manages explicitly. That's how it works for her, and whatever works is fine. Angela's model has certainly proven successful. We wonder, however, whether she would have had a more steadily paced pipeline if she had the resources of a discovery director, an incubation director, and an acceleration director rather than managing the entire portfolio herself. She, herself, believes that an even pacing is more palatable for the company. It's all about the resources that one has to make that happen.

Managing the Balance of Resources and Capability Development Across the DNA Activities

As we mentioned in Chapter Five, accelerating businesses often uncovers new opportunities. How can the system ensure that new learning that might initiate a new discovery is cycled back? Similarly, how can companies ensure that what's going on in discovery is fully mined? Corning's director of research constantly reviews on-the-shelf technologies to reconsider them in the light of the ever changing environment. Some are moved into incubated opportunities as feedforward actions. Finally, what happens when incubated projects are moved into acceleration and things don't go as planned? Are they killed, recycled into incubation, or added to one of the other portfolio programs?

A team in one of our companies developed a modulator for maintaining stability on vehicles under turning conditions. The product was initially commercialized for high-performance cars, but the vision was for a much larger mass market. When it came time to mass-market the device, additional development was required to drive costs down, engineer the device for various turning radii, and other technical barriers. So part of the venture team was moved from the business unit to an incubator to ensure they could fully leverage the technology platform without having the opportunity stalled for failing to meet operating unit expectations.

In another company, the project leader had explored many potential applications for the technology, but when it came time to accelerate the business, the newly appointed program manager did not leverage that learning. In a territorial tour de force, he began exploring new markets from scratch, many of which had been explored and discarded, slowing project acceleration. We illustrate this problem with Figure 6.1, where there is a disconnect between incubation and acceleration. This is a great example of why the DNA system requires an orchestrator: to monitor the system's linkage to the company's capacity and ensure that

Figure 6.1 Managing the Balance of Resources and Capability Development Across Discovery, Incubation, and Acceleration

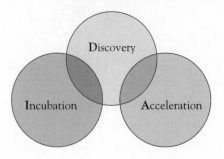

the system's interfaces are well greased, leading to continuous, enriched opportunity identification.

One can imagine many ways for the set of competencies to fail as a system. Most of these are due to a focus on one of the building bocks over the others, or what we call *system imbalances*. Sometimes there's a good reason, and sometimes there's not. The balance may need to be modified in accordance with your company's capacity. Vigilance on the part of the DNA system leadership is necessary to be aware of what the current balance is and consideration of what it should be. Following are some red flags of imbalances that we saw in our companies and their consequences.

Can't Get Heard. In this situation, lots of money is poured into discovery, but no one invests to explore the business opportunities. Many companies told us that they had lots and lots of great ideas, but could not draw attention to them. Most began building what we are now calling an incubation function, but no one was calling it that then: Kodak's additional personnel in the beta group of its System Concept Center, the growth of Sealed Air's corporate new business development function, MeadWestvaco's Innovation Process team, and Corning's exploratory marketing team all are

Figure 6.2 Can't Get Heard

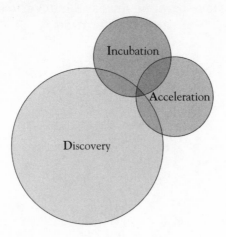

signs of ways to deal with the "can't get heard" frustration. We have portrayed this in Figure 6.2 as an overemphasis on discovery while neglecting to invest in incubation or acceleration.

Courage to Continue.   At one company, we were told that as the company came to understand the market opportunity, it didn't have the courage to continue to invest and execute on its learning: what everyone had learned was too foreign to them. So they did some incubation, but did not persist on the journey down the long and winding road. Great ideas were killed off after initial lackluster success, or incubated projects were killed when it appeared that the new business was not familiar enough. New business creation takes real courage. This company lacked conviction in the value it was bringing to the market. We portray this as Figure 6.3.

Big Ideas, Incrementally Executed.   In this imbalance, there is little incubation. Ideas become projects that move straight from discovery to acceleration. The company devotes lots of resources to discovery, but then commercializes only the most immediately apparent application and calls it a day. This is not new business creation; it's new product development. There's a big difference.

Figure 6.3 Courage to Continue

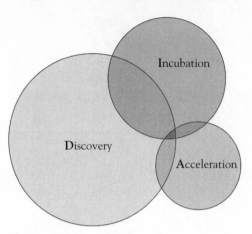

The CTO of a company in our sample that was experiencing this problem claims they are underleveraging their breakthrough business opportunities by simply going with the first, most immediately obvious application. They are doing this because the mandate calls for bigger hits, and when the first application is a bigger hit than what the business unit is used to with its incremental innovations, they're happy. So they're getting somewhat bigger ideas, but not incubating them as business platforms. This kind of mistake makes breakthrough innovation too expensive. The company is not achieving as high a return on its innovation investment as it could. We portray this as Figure 6.4.

Some of our firms developed useful practices to defend themselves against incrementalizing opportunities. GE's CEO program—so called because it is sponsored by GE's Chief Executive Officer—is designed to harvest immediate opportunities for the business unit, generating immediate cash and stimulating the market, while at the same time preserving the experimental aspect of the advanced technology platforms. They do this by ensuring that most of the team members stay with the long-term platform project team, adding people through the CEO program to commercialize

Figure 6.4 Big Ideas, Incrementally Executed

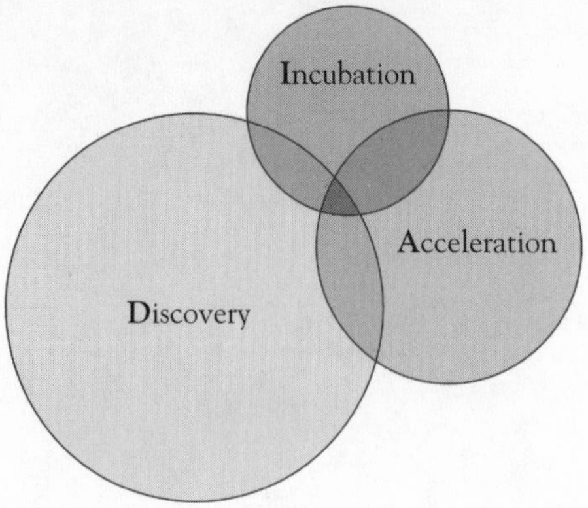

the short-term opportunities. IBM adds young talent who have just graduated from its Extreme Blue program to its graduating emerging business opportunity (EBO), or horizon 3, projects. It does this so that the Extreme Blue alumni, new employees fresh out of a training program devoted to new business creation, will continue to consider and actively seek new applications for the business platform.

Open Innovation at the Extreme. Many companies believe they can build breakthrough innovations by open innovation, that is, sourcing ideas, opportunities, and technologies from out-side: universities, small companies, venture capital syndicates. That may be true, but they cannot allow the discovery capabil-ity to shrivel. It's key to understanding real opportunity. Even if some companies do not spend a large portion of their budget on R&D (Johnson & Johnson Consumer Products is one example), the ability to evaluate technologies and elaborate them as business opportunities is mission critical. We portray this as Figure 6.5.

Cisco has been highly adept at open innovation over the past decade. It has used its inflated stock price to buy emerging tech-nologies and excelled at leveraging them. After the dot-com crash

Figure 6.5 Open Innovation at the Extreme

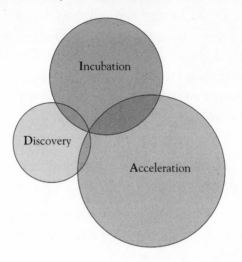

and Cisco's stock price plummeted, it no longer had the resources to buy new technologies and lacked the internal mechanisms to develop breakthroughs. We are mindful of the increasing emphasis on open innovation.[2] However, if a firm engaged in open innovation cannot seek out market opportunities for those technologies, whether the technologies originate inside or out of the firm, nothing will happen. Furthermore, once technologies and opportunities become identified, how will they be matured and brought to the market? The importance of incubation and acceleration is clear here.

Counterbalancing System Imbalances

System imbalances can be debilitating. The worst part of it is when a firm wants to build a breakthrough innovation capability but doesn't recognize the need for all three building blocks to be fully functioning. Each aspect of the DNA system has its own importance and place in the breakthrough innovation value chain. We've shown examples of the dysfunctions of the system and their consequences due to imbalances among the DNA elements. Again, we are not suggesting that all three building blocks must be resourced equally in order for the balance to be appropriate. An appropriate balance depends on each organization's capacity for innovation, and that, of course, changes over time. That's why an orchestrator and an innovation function leader are critically important.

Managing the Health of the Breakthrough Innovation Portfolio

The third system-level activity is BI portfolio management. By this, we do not mean managing the portfolio of all innovation projects within the company (incremental to breakthrough). For now we're talking about the set of projects within the BI system: all high risk, all highly uncertain, and all potentially big returns. For many companies, this will be a new way of thinking. Rather than considering BI projects as one-off opportunities, the mental model is one of a set of opportunities—the portfolio that can be managed.

This is portfolio management because we're not dealing with one lone project. We're building a corporate capability, meaning many opportunities are being worked at the same time. Trade-offs will have to be made, and resources (money, of course, but also attention) are limited. So we need to think in terms of a portfolio within which to make those decisions.

Those who are developing and managing a portfolio have many issues to consider: portfolio size, diversification, churn rate, pacing the portfolio (which we've already mentioned), and cross-portfolio management.

Portfolio Size

Most companies' BI portfolios are limited by a preset budget. If that is the case, the portfolio management team must consider how many projects should be in the portfolio and how much funding each should receive. Clearly those in incubation and, particularly, acceleration require many more resources than those in the early stages. GE was clear: six ATPs were identified (with one eventually added). Shell Chemicals funded nearly anyone who applied to the GameChangers program, but only a very small amount of money to keep them working during their free time. At MeadWestvaco, three or four businesses were being nurtured over nearly the entire study period; the portfolio size did not change. At IBM, the portfolio of corporate EBOs grew from ten to nearly thirty, at which point Bruce Harreld realized they could not continue to spend the necessary time with all those fledgling businesses and had to push some of them to horizon 2 status. The pipeline had swollen. As each of these examples attests, it's important to consider how many projects and how many resources per project in each of the discovery, incubation, and acceleration portfolios you are prepared to invest in. Granting too few resources to too many projects will ensure that none makes adequate progress, and funding only one or two at the expense of a diversity of opportunities is just too risky.

Portfolio Diversification

Typically when one thinks of portfolio management, one thinks of diversification. Traditionally we're talking about diversification to hedge against risk. That's not the situation here, since they're all highly new, highly uncertain opportunities. But diversification is still important, just along other dimensions. The BI system mandate is to help stimulate businesses for the company's future, so putting all of the eggs in one proverbial basket may not be wise. Most of our companies weren't thinking this way at the outset of the study in 2001. They were just trying to identify some project or programs and get started. That's fine, but eventually they have to consider the projects from a strategic perspective. We observed four possibilities for diversifying a BI portfolio.

First is along technology domains, platforms, or competencies. This was one of the approaches of Johnson & Johnson Consumer

Product and Albany International. The idea is to develop a variety of leading-edge technology competencies that can be leveraged quickly into opportunities as they are identified. The key is to constantly seek the opportunities to leverage these in the marketplace.

Second is along business or market domains. GE's CTO, Scott Donnelly, explained all of the six initial advanced technology platforms in terms of where the business opportunities were expected to be in ten to fifteen years. GE could see that biotechnology was going to be a big impact on the market, and GE Medical Systems was not big enough or prepared enough to handle that future. So they began a technology program to prepare themselves. As he explained to us regarding this and other ATPs, "It's a natural space for GE to be a leading player."

Third is by time horizon. A number of our participating companies wanted to ensure a steady stream of breakthroughs over time, so they looked for short-term programs to begin with, as well as some longer-term programs to fuel their pipeline. This diversification dimension is obviously more difficult to live by, since what appears to be an easy and obvious opportunity may shrivel and what appears to be far off may steamroll to the market.

Finally, companies considered diversification along the lines of organizational fit. In other words, they wanted to balance their portfolio in terms of businesses that would fit into ongoing operating units, those that required their own organizations, and those that could be drawn on from multiple organizations. While in the past we've seen companies attempt to balance businesses that were spun out of the corporation with those that are spun in, the former have fallen out of favor since the late 1990s, given the fact that most such spinouts did not benefit the mother corporation strategically. Sometimes, however, companies spin off fledgling businesses and then buy them back again as they gain momentum.

In all cases, the diversity of the portfolio needs to represent the company's strategic intent. Projects evolve as learning occurs, and so every once in awhile, a check must occur to make sure the fit is right. Diversification is great as long as it's the result of explicit decisions and not random.

Churn Rate

Venture capitalists talk about the churn rate in their portfolios of investments. By churn, they mean the rate of replacement of projects that are either killed or assigned a lower developmental priority and put aside for later development with new ones. Projects may meet dead ends or run into a currently insurmountable problem. Churn is expected in the world of high uncertainty. There's simply no way to predict the outcome of these opportunities prior to going down the learning path a bit.

Just like the venture capitalist's world, we expect a fairly high rate of change in the portfolio of breakthrough innovation opportunities as well because so many of them fail. Again, if there is no failure, there isn't enough attempted risk. In fact, many don't make it out of discovery. There are great ideas that won't work or don't live up to their initial promise. This happens all the time since people are stretching the boundaries of what's known in their attempts to bring new value propositions to the world.

The question is, What are your expectations for churn within each building block and across the entire DNA portfolio? You won't know at the outset, but you can monitor this over time and decide if you are comfortable with it. Only one of our companies thought about this and measured churn rate across discovery and incubation (not acceleration). At times they told us their churn rate in discovery was too high. The innovation portfolio manager wondered if the evaluation board was becoming hypercritical too early on. Traditionally venture capital investments yield about one wild success for every ten investments.[3] But that success pays for all the failures and then some. We'd expect that venture capital success rates wouldn't apply; large firms should do better.[4] After all, large companies have massive stocks of resources, rich networks, and a coherent strategic path. One would expect higher success from such organizations than from a venture capital portfolio invested in lots of independent entities that do not draw on one another or any other asset base other than what the individual ventures can access on their own. Those of our participating firms that were willing to consider the question told us they expected anywhere from 20 to 50 percent of their BI portfolio projects to achieve success. That's great in comparison to the venture investment market, but it is difficult for companies uncomfortable with risk. As one of our BI portfolio managers said, "We're just not used to looking at ten risky things knowing only two big ones are going to come out of it."

If your churn rate is uncomfortably high, there could be many reasons. Are they poor ideas? If your answer is yes, are people under too much pressure to generate ideas, such that creativity and opportunity elaboration aren't given the time they need to ignite? Are your projects making the transition too early, such that they fail in the next phase because they're too immature? Are your teams poor at executing on their plans? Are they following through, or are they stretched across too many projects and unable to devote adequate attention to this one? Or maybe you are populating teams with the wrong sorts of skills. How well are you paying attention to team composition? Do the teams have adequate resources to test the assumptions they're working under and learn? Are the projects evaluated using criteria that are too loose? If, for example, projects are so poorly fitted to the company's strategic intent that they'll be hampered for resources once they begin to gain traction, then you may be approving them now, only to kill them off later.

If your churn rate is too low, perhaps the portfolio teams are scared to take a real risk. Perhaps the evaluation board in discovery is sticking too close to what's already known. Perhaps projects are being allowed to linger without making steady progress. Portfolio churn rate is a revealing diagnostic to consider.

Pacing the Portfolio

It is easy for the portfolio manager or portfolio management team to get caught up in the projects in incubation and acceleration and forget to replenish the pipeline. This happened in several of our companies. The BI portfolio management team felt so pressured to get one or two projects graduated that they focused nearly all of their time on that objective. In the meantime, their idea hunters were finding new opportunities but could not attract the attention of senior leadership.

One way to handle this apparent distraction is to be less concerned with balance across the three building blocks and focus more on guiding a set of businesses through discovery, incubation, and acceleration as a cohort. This seems workable when one has flexibility in the use of resources and can redirect staff from one block to another if necessary. However, given the differences in skills and resources necessary for each building block, this seems unrealistic. The lumpiness of resources required may also stress the company's coffers, so theoretically, one would prefer a steadier pacing of BIs through the DNA system. Recognizing that there will be far fewer opportunities that are being accelerated than are initiated in the discovery function, due to the natural rates of failure associated with BI, the portfolio trimming should be taken into account in considering such pacing.

Cross-Portfolio Management

Finally, cross-portfolio issues are all major aspects of managing the BI portfolio: redundancies or convergence from independent programs, synergistic effects of one platform's learning with another or of one platform's learning with other ongoing businesses that were never previously considered as related to the initiative, and monitoring which projects should be starved and which should be fed based on their collective progress. As Gary Einhaus at Kodak told us, it's an enormous scope of responsibility. It also requires a great deal of both discipline and creativity.

In the mid-1990s a GE global research scientist who was dedicated to the aircraft engine division's needs came up with a way to capture movement with digital imaging technology. This was a critical technology for the aircraft engine division, but he also realized that the medical systems division might find it of value. He contacted several people within the medical systems business, as well as their counterparts in R&D, and eventually, GE's digital X-ray business resulted. Ensuring a rich capacity exists to enable the kind of networking required to uncover these synergies is all part of the DNA system leaders' responsibilities.

A Management System for Innovation

The familiar figure showing the elements of a management system appears (again) in Figure 6.6. While we have scrutinized how each of these elements is executed for discovery, for incubation, and for acceleration, none of these alone will make breakthroughs happen.

Figure 6.6 Management System Elements

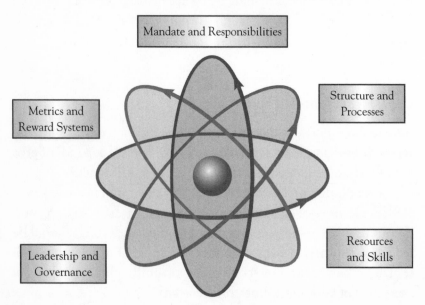

Let's consider DNA as a management system for innovation without belaboring the points that have already been hammered home.

Mandate and Responsibilities

The innovation system mandates are about the flow and pacing of breakthrough projects: ensuring a steady investment in and delivery of new businesses that renew the company by bringing step-change value to the marketplace.

Structure and Processes

Where should the DNA system reside in the company? Should each of the building blocks report to the same senior corporate officer? Should they all be under one umbrella, or does it make sense to split them up? It's amazing how much turmoil we observed in the organizational structures that companies developed to house their innovation function. But that's because they continuously evolved their level of sophistication and recognized the need for more of each building block over time. We offer several models that firms held on to after they experimented for some period of time.

Holistic, Sequential Model. This structure is what Kodak eventually evolved to, as we described in Chapter Five. Discovery, incubation, and acceleration are all housed under one umbrella organization (Figure 6.7). It's designed so that a project will pass from one group to another as it matures, but the groups are tightly linked to one another and report to one senior officer, typically the CTO, although governance boards of a broader constituency carry the decision-making authority.

This system optimizes the freedom of the portfolio businesses to develop unencumbered by business unit oversight, but they suffer from the same problem: lack of business unit oversight makes the transition of newly developed businesses into operating units very difficult. The use of these systems requires a strong governance board

Figure 6.7 Holistic Sequential Model

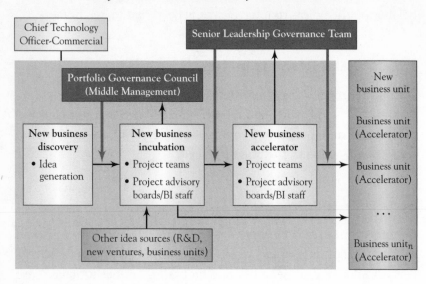

that can ensure that new businesses are heading in a direction that's exciting for the firm in some way, either as opportunities that won't ever end up in one of the current divisions or as opportunities that will, but will challenge the division to change in some way.

Self-Similar Model. A self-similar model is one whose BI infrastructure is modeled at the top level of the corporation but is mimicked throughout the rest of the company, as shown in Figure 6.8. This is the model IBM developed. So again, discovery, incubation, and acceleration are all overseen by one group (the strategy organization in IBM's case), but only multialigned and white space opportunities are overseen at the corporate level.

A similar structure is set up within each division for projects that appear to have potential impact that would be aligned with the division's current customers, business models, or technologies. This approach has the advantage of holding more people in the company responsible for far-future innovation, thereby changing the organization's culture toward innovation. But it requires persistence in ensuring that the divisions are doing their part in nurturing breakthrough opportunities, since the pressure for profit margins

Figure 6.8 Self-Similar Model

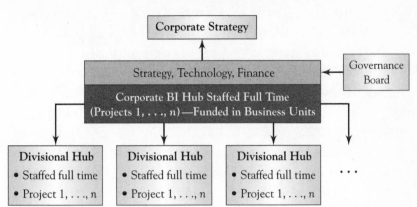

Figure 6.9 Mirrored Model

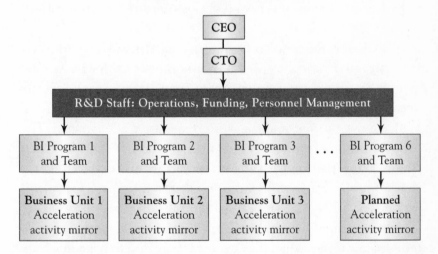

is always present. If this model is implemented, divisions should be measured on their success in commercializing breakthrough innovations, along with current term revenues and profits.

Mirrored Model. GE and Air Products have a very different model: development and part of incubation take place within R&D, and part of incubation and acceleration take place in the division (Figure 6.9). The distinctive aspect of this model is that they all take place at the same time for any single business opportunity. The projects are identified, selected, and incubated within or in close connection to the R&D organization. Simultaneously, in divisions that appear to be the ultimate appropriate home for a particular BI, a complementary acceleration activity is initiated even before there's anything close to a marketable product. A general manager is hired or appointed to begin building the business's infrastructure, including searching for potential acquisition candidates, value chain partners, and appropriate talent to bring into the organization. This approach seems most appropriate when the company is heavily committed to making the new business happen in one way or the other. They do not consider failure an option. GE, for example, identified six (and then a seventh) ATPs and was committed to their success in one form or another from the outset. This was not an options approach to innovation.

Concluding Remarks on Organizational Structures. Ultimately it doesn't matter if the three building blocks are located under one umbrella organization or separated, so long as each is aware of the others' roles, each is measured based on appropriate performance expectations, and their interfaces are well managed. The innovation function leader also needs to ensure that the DNA system is interacting with the organization, an issue we take up in Chapter Seven. The organizational structure that a firm elects depends on its culture and capacity for innovation.

One caveat: the major challenge that we see companies face in terms of location is when they expect incubation to happen within business units. It's extremely difficult to maintain the necessary patience and get attention needed from coaches in the incubation setting when they're located in business units. That's because the world of operational excellence dominates the business unit setting. That means the culture, the leadership, the metrics, the processes, the resources, and the decision-making systems are all set up for operational excellence, which is not a world of patience, learning and experimentation, and tolerance for failure. Remember that operational excellence is about variance reduction, while BI is about increasing variety to maximize learning. Each requires different mind-sets and management styles. So if you find that incubation is expected in your business units, either sound the alarm or ensure that the incubation management system elements are all present there. It'd be a subsystem within a business unit.

Innovation Processes. At a system level, the critical processes are diagnostic and monitoring processes, indeed, continuous improvement processes. Someone (the orchestrator and the innovation leader) must be vigilant, constantly auditing the DNA system to ensure it's working and politicking to get the repairs made to any breakdowns. Mike Giersch realized that not enough horizon 3 projects were making the transition to horizon 2, and so he raised the issue of the need to clarify the transition criteria. He later realized that the discovery pipeline wasn't filling, and so raised the issue of needing an additional staff person to help in discovery.

We wondered if Giersch ever sleeps at night. because it seems as if his brain is constantly on and constantly asking, "How could we be doing this better?" The same is true for Laura Pingle at MeadWestvaco; Nancy Sousa, and later Gary Einhaus, at Kodak; the vice president of engineering at Sealed Air; and Ed Hahn at Albany International. All of these people cannot stop diagnosing the innovation system. As they do, they develop tools to help them—tools to aid in the transition of businesses from discovery to incubation to acceleration to operations,[5] methods of developing career paths for the innovation function, tools to aid pipeline filling, and more. They're the orchestrators, working in conjunction with their senior leaders to constantly improve the innovation function. We deal with this in detail in the next chapter.

Resources and Skills

The innovation system needs an identified budget for the innovation function itself. Discovery, incubation, and acceleration need to be covered. This includes funding for projects, but also for support staff (coaches), prototype development, market exploration, and all the other activities we've described.

The good news is that the discovery activities (beyond the R&D budget) and much of incubation are relatively small investments and may be leveraged through the R&D budget that's already identified. Additional investment, however, is necessary in the later parts of incubation and acceleration. One of our companies, for example, told us that it was hampered in its attempts to develop breakthroughs because the investment required for prototyping fabrication processes and other potentially game-changing manufacturing methods was too high; indeed, a dedicated prototyping plant would be needed. The point is that if you're in the innovation game, it will take an investment of resources. Texas Instruments developed a whole new business unit and staffed it for the acceleration of its digital micro mirror device.

Although there are financial costs associated with later incubation and acceleration, the number of businesses that succeed to that point are fairly few. The much bigger cost, at least at the outset, is in the time and effort needed to build and maintain a culture of innovation. System leaders will find they need to communicate about the importance of and existence of the innovation function internally and externally. They may need to give the same speech thirty times in different venues to ensure that everyone knows that innovation is a key part of the business now. They'll need to explain to employees what the company's investment in building an innovation function means for them. GE had invested in the ATPs, to be sure, but also in many parallel programs in the business units, notably titled "imagination breakthroughs." This initiative has been trumpeted within the company and to the media relentlessly. Jeff Immelt is adamant about innovation, and he spends a substantial proportion of his time promoting it. Indeed, Immelt allocates 30 percent of his time to growth initiatives (and 30 percent to people evaluation and 30 percent to operations).[6]

The R&D budget as a percentage of sales for large firms is, on average, about 4 percent.[7] That's not a lot. And a limited amount of that budget is devoted to exploratory research, most of it spent on near-term business unit needs. The innovation system resources are needed to leverage that rather small investment in exploratory research and other externally sourced opportunities into substantial new businesses—breakthrough businesses.

Leadership and Governance

We've addressed responsibilities for portfolio-level issues, but the question is, Who should be responsible? Who should govern the DNA system? The idea of system-level governance is not to review projects but to attend to those in later stages due to heavy resource commitments to ensure the portfolio is healthy and the pipeline of activity is flourishing. The important aspect of DNA system governance is that it's not centralized under one person. Although the innovation function needs a leader, that person is not the sole decision maker regarding major BI investments. When autocratic decision making occurs, no one else feels responsible for the company's innovation agenda. We've seen many names for portfolio governance boards: Air Products' Growth Board, Sealed Air's Business Innovation Board, Corning's Technology Council and Growth and Strategy Council, DuPont's Apex Board and Growth Councils. All comprise high-level senior leaders of the company. Sometimes they get too big and cannot have true strategic conversations. Some of our participating companies told us that more than five or six people is too many.

When you think about DNA, it's important to ensure that the leaders of those building block competencies are on the board. In every case we've studied, discovery is "owned" by the chief technology officer, but in most cases, incubation and acceleration are not "owned" by anyone at the senior leadership level. The implications are exactly what we've reported to you: companies struggle to move breakthrough ideas ahead.

Metrics and Reward Systems

What does BI system success mean, and how would it be measured? Measuring the innovation system's performance is a must, but it's problematic for many companies, since they tend to measure innovation solely on financial terms.[8] There may be a long lag time between a potential breakthrough project's initiation and the point at which it develops significant revenues as a high growth business. Exhibit 6.1 lists some alternative ways to measure the health of the DNA system. Financial indicators are part of it, but only one part.

Exhibit 6.1 includes all of the elements of the portfolio's health, as we discussed earlier in this chapter. It also considers interface management between the three building blocks, including feedforward and feedback loops. In addition, one must consider the impact that the innovation portfolio is having in the marketplace in terms of adding to the company's reputation as well as bringing game-changing value propositions to the market. Finally, there are a number of considerations of the innovation system's impact on the company. One of them is financial—but only one. The rest help rejuvenate the company, renew its people, and create an invigorated organization ready to exploit every opportunity that arises.

Rewards and Punishments. One thing we have discovered in our more than ten years of research on the subject of breakthrough innovation is that people talk about providing financial incentives to those willing to risk their careers to develop breakthrough opportunities, but no one does it, for lots of good reasons. Differential financial incentives, like equity stake in the new business, cause divisiveness, we were told. When the team needs to draw on company resources, there's a propensity not to share the best resources if the profits won't also be shared. This, they tell us, breaks down innovation cultures. This makes sense. When companies were creating spin-out companies in the late 1990s (Lucent,

Exhibit 6.1 Measuring Your Innovation System's Health

Instructions: Circle the rating that best describes where your company is now.

	Exceeding Objectives	Consistent with Objectives	Not Meeting Objectives
Health and activity of the portfolio			
Portfolio size			
Number of projects	2	1	0
Average size of projects (potential)	2	1	0
Adequacy of resources per project	2	1	0
Number of new ideas	2	1	0
Number of new projects initiated	2	1	0
Portfolio diversity			
Diversity across technology domains	2	1	0
Diversity across market domains	2	1	0
Diversity across time horizons	2	1	0
Diversity across aligned and unaligned opportunities	2	1	0
Portfolio churn			
Overall portfolio churn rate	2	1	0
Churn rate in discovery	2	1	0
Churn rate in incubation	2	1	0
Churn rate in acceleration	2	1	0
Portfolio pacing			
Estimated time to commercialization across projects is well balanced	2	1	0
Number of projects making the transition between stages	2	1	0

(Continued)

Exhibit 6.1 *(Continued)*

	Exceeding Objectives	Consistent with Objectives	Not Meeting Objectives
Cross-portfolio health			
Synergistic effects across projects	2	1	0
Redundancies across projects	2	1	0
Interface management			
Number of projects making the transition between stages	2	1	0
Smoothness of handoffs from discovery, incubation, and acceleration	2	1	0
Communication flows from incubation and acceleration back to discovery as new opportunities emerge	2	1	0
Impact on the market			
The company is gaining external recognition as an expert in a particular technology domain	2	1	0
Richness and promise of projects: potential to be game changers	2	1	0
Impact on company			
Number of projects that make the transition to business units or other permanent home	2	1	0
Financial impact of those projects on the company	2	1	0
Impact of learning within projects on other business arenas	2	1	0
Spillover of BI management system elements to other high-uncertainty arenas	2	1	0

	Exceeding Objectives	Consistent with Objectives	Not Meeting Objectives
Development of entrepreneurial talent within company as project leaders are recycled and moved to business units	2	1	0
Increased confidence in the company's ability to innovate	2	1	0
Increased robustness of new ideas coming in.	2	1	0

Nortel Networks), they did use this model. However, that worked for ventures that weren't making the transition into the company and, in fact, served to widen the separation between the new business and the mother ship. That's exactly what firms trying to develop a BI capability want to avoid.

Our participating companies told us that project team members benefited from the opportunity to present to and work with senior leaders in the organization. Their visibility was increased, which enhances the opportunity for promotion over time. In fact, several of our companies are now requiring a rotation through the innovation group for individuals viewed as high-potential employees who will one day be members of the ranks of general management. We certainly agree with that approach.

Most project team leaders and members are more concerned with protection from punishment when projects fail than with differential rewards for projected successes. It's more important to remove disincentives than to give incentives. Individuals can get burned in this system. Here are a few testimonials to that fact from the people we interviewed:

> "Not only is there a question of 'can,' there's a question of 'why' would anybody want to [be part of a radical

innovation project]. Because I've never seen people like that get rewarded in the past. They're sort of outcasts."

"There are a lot of perceptions of people's careers stalling or actually ending because they undertook something very innovative or radical within the organization."

"I think that's how sometimes these projects go on for a very long period of time because people stake their reputations on it. People in some cases believe they stake their careers on it. People feel, 'If I kill the project, I'm killing my credibility.'"

"I think if you have the right people, the right discipline, you don't need all this other motivational stuff, because people get very focused on the goals."

People with innovative skills are driven to succeed. The role of the DNA system overseers is to ensure they have the freedom to do their job.

Conclusion

Simply having effective discovery, incubation, and acceleration capabilities is not enough for a useful breakthrough innovation system. The parts need to coordinate and communicate with each other, and they should operate in some sort of balance, in accordance with the company's strategy and capacity for innovation at the time. This chapter is about how the DNA parts should work together and the challenges for having that happen. The DNA system oversees and nurtures a portfolio of initiatives, diversified along many dimensions. The innovation system's health is reflected in the success of the portfolio of breakthroughs, not its individual breakthrough business initiatives, since they're all fraught with risk and uncertainty.

Now we need to tackle the task of knitting the innovation system with the rest of the company. That's the role of the orchestrator.

Questions for You

1. What are your biggest challenges in making the system work?

2. What is the primary interface challenge you face?

3. Are there systemic incentives for people to join or advance BI projects in your firm? Disincentives?

4. How balanced is your system? What are the implications of any imbalance you may have? Is balance always important?

5. What senior-level executive, if any, is responsible for incubation of new businesses in your company? For acceleration? Are these the appropriate people for the role?

7

INCORPORATING THE DNA

The Role of the Orchestrator

Up to this point we've considered the discovery, incubation, and acceleration building blocks and how they work together as a system. We've talked about portfolios, transitions, pacing, balancing, and all of the issues that ensure or hinder the steady progress of breakthroughs as they advance toward maturity. Now we turn our attention to this reality: you have built a system, it's working, and you think it's running smoothly, but your organization either cannot absorb what you're producing or wants much more than you've got. This mismatch occurs because of the company's ever changing capacity, as we discussed in Chapter Two. One thing's for sure: capacity levels are not constant. Sometimes the company is in love with innovation, and sometimes it's not. That's why we need to orchestrate the relationship between the DNA system and the mainstream organization.

We've already addressed this in many ways. When we describe governance councils for discovery, incubation, and acceleration, they've typically incorporated senior leaders from divisions as well

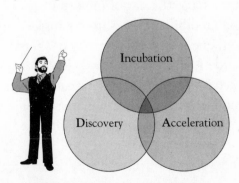

as those in the corporate-level ranks who concern themselves with the company's future health. These are critical mechanisms for orchestrating the relationship of the innovation function with the rest of the operation. But there's more we can, and need, to say on the subject.

Orchestration is the activity of managing the DNA building block system in a manner appropriate for the organization's capacity at any given time. This includes times of organizational stress (the profit picture isn't healthy, new competitors are on the scene, legal issues are bogging you down), as well as times of organizational munificence (sales are up, margins are great, new product launches are hitting big). In recognition of ever changing capacity, the orchestrator must alter the pacing, the funding levels, and sometimes even the scope of the innovation function. Our point is to adjust this rather than to completely shut it off, as companies tend to do. Innovation must be a permanent fixture, not a program or a phase, in the organization's spectrum of activities. Otherwise the start-up costs and the shut-down costs are too high. Orchestrators think about innovation system-level problems. They constantly audit the DNA system. They consider the logic and gaps of the system and make changes to improve it. They ensure that the management system elements—talent development, resource availability, appropriate processes, and metrics—are aligned with organizational expectations for breakthrough innovation. They also ensure that organizational expectations for breakthrough innovation are realistic.

Orchestrators have the ear of important people. They have influence. Their credibility comes from a history of solving company issues in unique ways with a variety of approaches, or from a long track record of building businesses in other companies they may have worked for previously in their careers. They push senior management and help them to articulate strategic intent for the company. They clarify roles and responsibilities of the BI system in relation to other parts of the company and other groups engaged in new product development. They clarify the role for BI within the larger innovation portfolio and monitor that it's not being squeezed or, alternatively, relied on too heavily. They clarify portfolio objectives across DNA and monitor the pacing of the opportunities in accordance with the organization's capacity at the time. They oversee decisions regarding the landing zones for new businesses graduating from the BI system. They broker connections between the innovation system and the mainstream organization or even beyond it, to the larger marketplace, in the interest of specific projects that need particular types of expertise or resources. Finally, they work to ensure that the mother organization is aware of the BI system's impact on it. It's a big job, and it's very strategic.

We encountered a number of orchestrators in our research. Ron Cotterman at Sealed Air, Ron Pierantozzi at Air Products, Mike Giersch at IBM, Mark Newhouse at Corning, Rick Spedden at MeadWestvaco, Dave Austgen at Shell Chemicals, and there were many others. Some are vice presidents, others directors. The orchestrator must have the status and power to understand and act on the strategic landscape of the company. In different companies, these people may reside in different levels of the organization, but generally orchestrators cannot be at the very top of the organization or too far down. They need a bird's-eye view of the innovation system so they can fine-tune it—a view of the whole innovation system and how the pieces fit together, and how it interacts with the larger complex mother ship. They're organizational stewards. They care about the health of the company.

Orchestrators in our study described their role as "not rocket science but very complex." They have to deal with a very wide range of issues simultaneously, more than they'd experienced in any other job in their careers. They need to maintain relationships with a large web of people. Obviously they must be politically astute.

The Orchestrator's Agenda

Orchestrating occurs within the DNA system. Issues of the transition of projects, pacing projects within the pipeline, and portfolio composition must all be orchestrated in accordance with the company's capacity at the time. In this chapter, we examine issues associated with orchestration between the BI system and the mainstream organization given the high likelihood that the company's capacity for breakthrough innovation changes often:

- Managing the breakthrough innovation function's perceived role in the firm
- Monitoring mandate creep
- Orchestrating transitions of BI businesses out of the innovation function and into landing zones
- Orchestrating to get things done
- Orchestrating linkages to company leadership
- Orchestrating linkages to other corporate functions
- Orchestrating uplift of the innovation function

Managing the Breakthrough Innovation Function's Perceived Role

Most companies engage in many types and levels of innovation. Continuous improvement of current products and services, incremental changes to current product lines, platform-level evolution, and breakthrough innovation are all part of the mix. In addition, there are considerations regarding the relatedness of the innovation to current product and business domains. Some are completely aligned, and other investments may be more diversification efforts.

Finally, how tightly do you want to hold to the concept of strategic fit? How much do you plan for, and how much serendipity is allowed? Case studies of breakthroughs repeatedly point to the role of luck and opportunism in many instances. Here is one famous example:

The famous story of DuPont's Surlyn® demonstrates how luck, opportunism combined with commitment, and patience of more than one person can eventually lead to a winning business.[1] Richard W. Rees, a research scientist at DuPont, was conducting a scouting project where he had few constraints on where and how to explore. To begin investigating one dysfunctional reagent, he had to prepare a chemical salt of the plastic he was testing. Much to his surprise, an unusual gel formed with an exciting combination of potentially useful properties. The development of this tough, stiff, resilient, bouncy, glassy, clear plastic that he created was pushed through the organization. A commercial facility was authorized a year after its discovery, and by 1963 a pilot plant was completed. DuPont announced the new materials as Surlyn® resins in the September 1964 issue of *Modern Plastics*, and samples were sent everywhere that toughness and clarity were demanded.

But the development effort ran into big problems. During some early molding runs, for example, this amazing new adhesive glued the molds together so firmly they could not be taken apart. In 1965, a 30-million-pound-per-year plant was up and running without a single customer. Then the first customer was found: a producer of the tips of women's shoe heels in England—not exactly the high-volume application envisioned. Shortly after, a DuPont development person began working with RAM, a company with a very small market share in golf balls. Only after the "click test" (listening for the right sound while driving the ball in the early morning in a shady spot with the right amount of dew) did the coating pass. Given all DuPont's fancy laboratories and research personnel, what a letdown that the go–no go test depended on such a low-tech criterion. Now nearly every golf ball in the United States has a Surlyn® coating.

In 1969–1970 the development team came up with films for packaging processed meat, but only the smaller packers would work with them. Working with these smaller meat packers, DuPont was able to demonstrate proof of savings, and one packaging application led to another. By 1972, nine years after the laboratory discovery, Surlyn® reached break-even. Scientists were learning more about the material and about possible markets. They were able to tailor ionomers for a wide variety of applications. The product went into bowling pins, auto bumper guards, and ski boots, among other things. Markets grew that were never imagined, and now the science followed the markets. Today there are still untapped markets for Surlyn®. Here is a case where a technology sought markets, and markets found underappreciated properties in the technology. Together this translated into a winning business.

The point is that a breakthrough innovation capability is not the sum total of a firm's innovation efforts. The orchestrator must constantly remind the rest of the organization's leadership of the role that breakthrough activities play in relation to the rest of the innovation system and ongoing operations.

This allows the orchestrator and the innovation function senior leadership to continuously remind senior and business innovation leadership about what is realistic to expect of the breakthrough innovation system and to help define its boundaries in terms of time horizon, alignment with current businesses, and technology and business domains compared with other parts of the company's innovation engine. We heard many objectives imposed on the innovation function by others who had no clue as to what can and cannot be expected of breakthrough innovation. "We need to increase our price-to-earnings ratio in the next six to twelve months," we were told by one director when we asked him what he was hoping the breakthrough innovation initiative would do for the company. Clearly that's not going to happen in the time frame he expects. The orchestrator in that company had a lot of work to do to educate others about the realities of breakthrough innovation. It's about longer time frames, steady investment, and relatively low probabilities of success on any single project, but long-run, high returns over time and across a portfolio of projects.

Figure 7.1 provides an approach to thinking about the role of the breakthrough innovation in a company. One can imagine using this figure as a tool for planning and then for monitoring the efficacy of the company's innovation strategy. The numbers in the cells are not suggestions for what any company should do; they're simply an illustration. Each company has its own innovation strategy. Working through an exercise like this can help company

Figure 7.1 Developing an Innovation Strategy

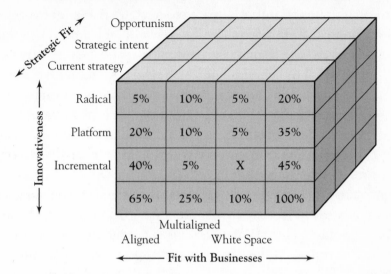

Multialigned
Aligned   White Space
◄——— Fit with Businesses ———►

leaders clarify their innovation strategy among themselves and then allocate their budget and monitor results accordingly.

Corning's strategy and leadership team identifies three buckets for R&D spending, labeled "today," "tomorrow," and "beyond." They work to ensure some level of activity is occurring in each of those and that the "beyond" projects don't get squeezed or completely quashed when the "today" projects are struggling. But companies can become much more sophisticated about this in terms of the buckets they identify, the allocation of expenditures across them, and the monitoring of performance across the multiple categories of innovation. The orchestrator's role is to ensure that the breakthrough innovation category receives a continuous flow of investment and attention within the entire portfolio of the company's innovation efforts.

Monitoring Mandate Creep

Mandate creep is the idea that innovation systems tend, under pressure, to gravitate toward more aligned or near-term opportunities. Conditions do change, and so do senior personnel. All organizational initiatives are predicated on a set of expectations grounded in internal and external conditions. As those conditions change, so must the expectations and, hence, activities associated with the system. This can be okay as long as it's clarified within the rest of the innovation system and the orchestrator and senior leaders are aware and in agreement. In fact, we would not call that "mandate creep" as much as a "modification to the mandate." The difference is about formality. In the former, the projects that are chosen may actually be out of alignment with the BI system's formal mandate, because the projects are selected for convenience or in response to pressures to perform. In the latter case, the breakthrough innovation mandate is revised to reflect the changes, so that the projects in the system are in alignment with condition changes. This becomes critical when the BI system's performance is measured against the firm's expectations of it. Mandates may need to be revised in accordance with changes in the company's capacity. This certainly occurred at Kodak.

When Kodak's Systems Concept Center (SCC) was initiated in 1994, it was given the mandate to find systems-level innovations that drew from across the company's current lines of business to become new growth platforms. But by 1997, that mandate evolved to breakthrough opportunities that were aligned with either current businesses or white space opportunities, since by that time it had become clear that those system-level, multialigned opportunities were few and far between and that some of the business units needed help in creating long-run opportunities for themselves. So over time, the SCC's chosen mandate incorporated aligned opportunities that were far enough in the future that the business units would not invest in them on their own.

There are other ways the innovation function can stray from its mandate that orchestrators must monitor:

- What is the right number of projects in discovery? In incubation? In acceleration? Is the system active enough? Too active? Are there enough projects in the portfolio or too few?

- Is the balance between organic growth and acquisition in alignment with the company's objectives? Has the company swung too far in either direction, relying on one method for growth too heavily?
- Are the opportunities big enough? Are they single products or platforms that could stimulate whole new lines of business?
- Are the portfolio businesses breakthrough enough?
- Is the BI system generating and nurturing its own portfolio, or behaving too much like a service organization for other units in the company?

Many of our participating company BI groups told us they spent some amount of time fielding calls from business units to solve their own new product development (NPD) problems. The BI group is acting as a consultant. Many relayed to us that it's tempting to do this to stay busy and gain credibility with the business units, but if they're not measured on these consulting jobs, and only on portfolio health, they are not really getting their job done.

Another activity that BI groups engage in is educating others in the company and helping facilitate idea generation and coaching workshops within the divisions. Again, this is important in many respects, but if it siphons too many resources and too much attention from the innovation portfolio, that pipeline is sure to suffer, with sad consequences.

The orchestrator monitors the innovation system's conformance to its mandate on all of these dimensions, and then alerts the innovation leader and governance boards so they can take action.

Orchestrating Transitions of Project to Operating Units

One major concern of the orchestrator is engineering the transition of projects between discovery, incubation, and acceleration; another is the transition of matured projects into the mainstream organization. We should not underestimate the difficulty of timely transition of BI projects to operating units. Some of the difficulty in the past has been that transitions have not been proactively managed. The orchestrator must ensure that management oversight occurs; he or she can work with the innovation system leader to assign a transition oversight team to facilitate moving breakthrough businesses from the innovation function and smooth entry into their new division or operating unit. The orchestrator's role is to ensure that a transition team is designated, comprising representatives from the receiving unit as well as from the innovation function, to oversee a smooth assimilation of the new business into the mainstream organization. We've related stories of breakthrough opportunities that were not given enough attention in their operating unit homes and died.

Other scenarios the orchestrator must guard against are the pressures of early transition because the first commercial application of a breakthrough appears so exciting, but there's no infrastructure to ensure that follow-on applications will be pursued. A negotiation between the innovation team and the receiving unit has to be undertaken to make sure that any remaining incubation and acceleration work is identified and responsibility assigned for its completion. In addition, transfer of people along with the business is a major concern. None of it will occur on its own. Transition oversight must be managed, and the orchestrator's job is to ensure it gets done.[2]

Orchestrating to Get Things Done

One point that emerges as we analyzed companies' various approaches to transition and funding is this: the stick approach works better than the carrot approach for breakthrough innovation. Companies that offered rewards and recognition for those engaging in innovation were, on average, less successful at developing an ongoing sustainable innovation capability than companies that levied negative consequences on those who failed to comply.

At the individual level, we've already mentioned that removal of punishment for trying was a better help than promising financial incentives. Another step along this line of thinking is to impose pressure on those who are not trying to innovate. This played out in a number of ways in our companies:

• *Funding oversight*. Some companies identify a specific set of accounts or funds for the breakthrough innovation projects and then monitor them to see that they're actually being spent. At IBM, for example, these funds are part of the divisional budget rather than the corporate budget. The corporate controller has visibility into the expenditure rate of those funds, and if the spending rate slows, indicating a lack of investment in (and therefore a lack of progress on) the innovation portfolio projects at the division level, the senior vice president of strategy contacts the division to find out why that investment is not being made.

• *Price premiums*. One company in our sample charges business units for projects with no risk that they are expected to adopt and subsequently commercialize. That's not so bad, since it's a lot like acquiring a young company, and the business units have experience with acquisitions. However, if the business unit hesitates and elects not to adopt the fledgling opportunity, the price goes up to reflect the fact that the corporate innovation group must continue to nurture it in its own portfolio and will continue to lower the risks for the opportunity. As risk goes down, price goes up. Eventually the business unit must acquire the fledgling business, so it's a matter of how much of a premium it is willing to pay for stalling.

• *Answering to the CEO*. In several of the companies we studied, innovation is front and center on the CEO's agenda. Although this may not always be beneficial for the organization (she or he may become overinvolved in the discovery and incubation phases, creating too much time pressure, and perhaps even micromanaging the projects), the mere fact that innovation is important to him or her is a signal to the company, and that signal can be manifested in terms of fear. "I'd hate to have to go before the CEO and explain why we haven't paid attention to this" was a comment we heard in one of those companies in our study.

- *Change the Business Unit Charter.* One company reported that one of its business units was stalling on adopting a breakthrough business because it did not neatly fit the unit's scope, business model, or customer set. Senior management immediately changed the unit's charter to incorporate a wider scope of markets that the new business could address. So rather than offering the business unit incentives to try to make the new business a success, with a recognition that the fit was a challenge, it set up a situation in which the business unit had to comply.

All of these examples are true, and they all point to the fact that if innovation is part of the company's agenda and it is a stable, long-term commitment, everyone is expected to engage in their part of the agenda. If they do not, they will face consequences that are uncomfortable. This may go against the grain of many management theories and practices today. In our experience, though, it works, and it helps make innovation a sustained organizational capability. The orchestrator must find ways to sustain it. These examples provide some methods for signaling to the larger organization that breakthrough innovation requires everyone's cooperation.

Orchestrating Linkages to Company Leadership

One of the critical jobs of the orchestrator is to educate the company's leadership about the nature of breakthrough innovation and what it requires. We were struck by the amount of reading and level of awareness about management of innovation that many of the orchestrators had. Some even read academic journals on innovation management.[3] Most had read the recent popular books on the topic, and many had attended seminars on managing breakthrough innovation.

It's good they did. Most had lots of work to do to educate senior leadership. Senior leaders are obviously very smart people. But the truth is that most arrive at their positions because they've performed very well in mainstream organizations. That means they understand the culture of operational excellence but may find the highly uncertain world of breakthrough innovation unfamiliar and uncomfortable. Or they are so naive about it that they believe it's easier than it is. Table 7.1 shows some of the mismatches we observed in senior leadership's expectations and the realities of breakthrough innovation. We're sure there are others, but these give you a sense of the need for education.

In some companies, senior leadership says they're committed to innovation, but when it's time to invest, they just will not do it. One of our orchestrators admitted to us, "Our leadership is committed to BI intellectually but not in their gut." He expressed concern that the governance board would get bored with overseeing the long development cycles that typify real breakthroughs in their industry.

In other cases, leadership may be committed to innovation but doesn't understand what it takes and needs reminding. One of our orchestrators told us that he's connected through his personal digital assistant to the company CEO at all times and has a running conversation with him throughout the day, at the CEO's request, reminding him of the differences between an innovation management system and what it takes to run daily operations. In another, the CTO at one of the divisional levels, who'd successfully incubated and commercialized several breakthroughs in the transportation industry, brought in an innovation expert to explain to the corporate vice president for technology how to measure performance for high-uncertainty innovation projects. The divisional CTO supplemented the expert's presentation with examples from the division's successful projects, so the senior vice president could apply the external expert's frameworks to the company's experiences with innovation.

Table 7.1 Mismatches Between Senior Management's Expectations and BI Realities

	Management's Expectations	BI Realities
The impossible mandate trap	Identify a huge new business for a business unit and use their business development resources to do it, even though they're under severe market pressure on margins and corporate pressure to generate immediate sales.	Business units are typically measured on revenue growth and profit margin. Investment to develop a new potentially breakthrough business, which requires lots of new learning, new relationships, and new resources, is unrealistic, particularly when the business unit is facing pressure to perform in its conventional business lines.
	Give us breakthroughs that will raise our profit-to-earnings ratio within six to twelve months.	The time frame is highly unrealistic.
	Develop a portfolio of unaligned opportunities with your current technical and business talent.	It's difficult to assemble the same set of people and expect them to yield dramatically different results than they have in the past. New types of talent, new networks, and different competency sets are needed.
The "it's no different" trap	Give us breakthroughs using our current management systems.	For such high-uncertainty endeavors, the organization needs a different approach.
	Give us breakthroughs without educating anyone as to what to expect the investment requirements and success rate will be.	Unless the people who are deciding on the investment amounts are aware of breakthrough innovation time lines and typical levels of investment, amounts may be too small (or too large) and inappropriately paced.
The history trap	Give us specialty downstream value-added business model breakthroughs even though we've historically been a commodities supplier.	Every company has a history of how it competes. To expect breakthrough innovations in completely different domains in areas in which it has no history, no competency, and no networks is unrealistic.

Orchestrating Linkages to Other Corporate Functions

Historically project champions for breakthroughs broke rules to get the job done. They found ways to get around the purchasing group, the human resource group, the legal group. We knew one such maverick who negotiated his own partnership agreements and called the legal staff after the deal was essentially complete. Others find senior leaders who provide protection to allow the rule breaking to occur. But the advent of the orchestrator allows something more civilized to occur. We observe concerted efforts in many of our participating companies to link to other functions and require that they support the innovation effort rather than undermine it with bureaucracy.

The orchestrators in our companies told us that they try to find someone in each of the functional groups who can understand the innovation imperative for the firm and help devise solutions to the BI group's requirements. That person then champions the modified rules or processes within his own function on behalf of the innovation group. One of them told us that his goal was to have everyone working on the company's innovation agenda: "I want to leverage my staff a thousand to one," he told us. That meant that for every member of the innovation group, one thousand others in the company were working at least some amount on the innovation portfolio's agenda. That may mean human resources is developing recruiting and promotion strategies for the innovation staff, or the legal department is working out a new intellectual property policy, or the purchasing department is setting a budget for fast-track purchases of equipment needed by innovation portfolio teams. In one case, the orchestrator convinced the finance function to fund projects on a multiyear basis, which saved untold time and energy during budget planning time. The challenge that orchestrators tell us they face in this activity is that the turnover in other functions becomes a setback. There is a need for continuing education regarding the BI system's role in the company.

Orchestrating Uplift of the Innovation Function

Many of the initiatives we observed at our participating companies started out as middle-manager-led groups that were small and somewhat isolated. Over time, the initiatives gained visibility, and the innovation capabilities that were developed were imported to higher-level groups and had wider-ranging influence in the organization. The Systems Concept Center at Kodak was ultimately dissolved, but the principles, practices, and many of the people who had staffed the center were brought into R&D when Gary Einhaus returned there from his assignment as strategic product group general manager and chief technical officer in Kodak's Professional Photography business. At 3M, the corporate enterprise development group that was originally a small incubator group in R&D morphed and was eventually disbanded. But what was added was a business development team, run by Michele Nelson, that supports all of corporate R&D's efforts to build a true discovery and incubation capability. At Sealed Air, the technology identification process (TIP) team that was resident in the Cryovac Division on the merger of Cryovac and Sealed Air in 1998 was combined with CEO Bill Hickey's personally sponsored mavericks. This new group worked with the vice president of engineering in her new position as vice president of new business development and her associated staff. Finally, at Air Products, the growth board was reconstituted from a large group of divisional executives who reviewed every project, to a smaller board (six members) of corporate-level officers who considered developing entire platforms for future growth. What is the common thread here? In all of these cases, there was one consistent orchestrator or a series or orchestrators who knew one another and held a shared vision of what the BI capability needed to look like in the organization. Orchestration of uplift in the organization is exciting and demonstrates that the journey toward building a breakthrough innovation capability doesn't necessarily have to start from the top of the organization; it can start with middle management and spiral upward. Moreover, consistency of orchestrator or smooth handoffs from one to the other is important.

Orchestrating Under Different Organizational Contexts

It's obvious that the orchestrator has plenty to do. Now we add a complication: organizational capacity. Organizations differ in the degree to which the DNA building blocks are well established. We turn now to considering these varying contexts.

Orchestrating Under Stressed Capacity

Stressed organizations tend to exhibit mandate creep toward aligned opportunities. Links to the business units are tightened as the pressure of immediate financial returns overwhelms the company. In this case, it's critical to ensure that the DNA governance board maintains a corporate perspective. Business unit leaders sitting on the board are more likely to be interested in what the innovation system can do to help their unit-level interests than the larger organization's higher-order goals of growth, which may require reallocating resources away from their business unit and toward wholly new business platforms.

Orchestrating Under Munificent Capacity

When a company has a very rich capacity for innovation, there are still challenges; they're just different. The key challenge that our companies displayed is difficulty in keeping attention on the matter of innovation. After all, the company isn't feeling much pain at that point. The long and winding road of incubation can feel like a waste of time and money. Many companies told us that their BI systems are most vulnerable when one or two of their core businesses are performing well. It's ironic given the fact that under stressed capacity, the organization does not have the patient resources available to develop BI. The fact of the matter is that this is the time to be developing unaligned businesses opportunities. One orchestrator indicated that he has much more latitude to do so under a rich capacity situation, but when we observed the company's actual decisions, we saw that they changed approaches, changed priority projects, and even changed their innovation infrastructure over and over. So although they said they wanted to build a breakthrough innovation capacity and get some new growth, the sense of urgency wasn't there. This company has come quite a ways in developing its infrastructure, but in a much longer period of time than others in our sample.

We observed three ways that innovation is being institutionalized in companies today. In all cases, there is an identified group responsible for helping BI projects move along. They differ in the degree to which they have control or ownership over the projects. The best model for institutionalizing innovation in your company will depend on its size, its culture, its stress level, and other elements of its capacity.

Innovation Process Facilitators. This approach is the smallest departure from current practice. Innovation process facilitators serve educational and coaching roles. They educate regarding the differences between breakthrough innovation and conventional new product development. They help teams understand and follow management approaches appropriate for high-uncertainty projects. They meet with project teams once or twice to follow up on the educational objectives but are not regularly involved as coaches would be on issues of strategic importance. Innovation process facilitators do not have control of project resources or the ability to influence senior leadership. In addition, this approach does not help with team-level performance evaluations or portfolio-level issues. They are in place to help guide individual teams through the bumpy terrain of breakthrough-related issues. In one of the companies we studied, an initial attempt at developing a BI group failed when the CEO decided to take a more outward-focused venturing approach. Some of the initiators of the BI group moved into this function, working with business units to continue to educate about innovation, generate ideas, and help them articulate those ideas into project opportunities. That person told us, "We're not ready for breakthrough innovation. We have so much work to do to get this company to think strategically about innovation at all!" She's biding her time. Organizational capacity is too low.

Breakthrough Innovation Project Support Function. Typically labeled "new business development" or "market analysis," this is a group that helps with strategy development and incubation of individual projects as needed. They do not have portfolio-level responsibilities and are typically resident in or near R&D. They may help build the connection between the project opportunity and the appropriate business unit if one exists; if no business unit is appropriate, this group may be responsible for incubating and potentially accelerating an opportunity themselves. But in terms of considering how to generate new ideas, or the health of the portfolio, or whether discovery, incubation, and acceleration functions are structured and operating well, that's not what concerns them. They're focused at the project level.

Breakthrough Innovation Function. A full-fledged innovation function is a corporate-level function with reaches into a company's divisions if the firm is large enough. Just as marketing groups consist of market research, marketing communications, sales organizations, and perhaps others, so too does the innovation group encompass multiple responsibilities. A BI function is responsible for building and nurturing a portfolio of opportunities; for overseeing the health of the discovery, incubation, and acceleration activities; and for orchestrating the relationship with the mainstream organization, with the understanding that that relationship will constantly change. We observed this model, in various stages of development, in seven of the twelve companies in our phase II group, but it's by far the most significantly different approach from conventional practice. It's this model that we've fleshed out in detail in these chapters and summarize in Table 9.1 in the final chapter. Once you've become familiar with the principles of an innovation function, you can adapt the practices to fit your own organizational capacity.

Considering Capacity and DNA Capability Levels Simultaneously

Table 7.2 shows four possible situations in which orchestrators may find themselves. One dimension is the current capacity of the organization: stressed or constrained versus munificent or rich, and developed in detail in Chapter Two. The second dimension is the degree to which the DNA building blocks are well developed. Is the organization able to generate and handle innovation opportunities? Are the discovery, incubation, and acceleration capabilities finely honed? If you're starting from scratch, there's a learning curve to endure. But if the DNA management system is in place and functional, it can be tapped appropriately, given the capacity the organization has for innovation.

The top right cell of Table 7.2 is the situation all orchestrators would like to face: the company has a rich capacity and enough experience at this that it has developed the sort of management systems we've been talking about: discovery, incubation, and

Table 7.2 Relationship of Capacity to Level of DNA Competency

Organizational Capacity	Low DNA Competency	High DNA Competency
Rich	Champions; many simultaneous experiments	Management systems
Constrained	Educate others about innovation Middle management initiatives Get one or two projects started and claim success Hope for change or seek job elsewhere	Influence senior leadership Delay most unaligned opportunities Refocus the BI group's mandate Facilitate innovation processes throughout the organization Slow the pacing of projects through the DNA system

acceleration are highly functional, with appropriate leadership, mandate, skills, processes, performance metrics, governance, and resources, and they're well integrated with one another. This is a highly functioning breakthrough innovation capability.

The top left cell of Table 7.2 indicates a range of activities the orchestrator can do under rich capacity, but the DNA building blocks are not well developed. This is the most frequent case we encountered at the beginning of our study. All of our companies wanted to be able to commercialize breakthrough businesses, but they didn't know where to begin. This is the scenario in which champions thrive, since senior leaders are eager to enable them. In addition, the company may try a number of experiments to improve their success record with breakthroughs. We witnessed this specifically at Johnson & Johnson Consumer Products, where Neal Matheson, the CTO, issued a mandate to each of the twelve vice presidents in the company to try some sort of experiment that would bring breakthroughs about. One tried to start an incubator. Another reorganized her R&D staff to work on platform technologies. Others started a ventures group. Matheson was watching to see which approaches worked and which did not. It was a learning experience, to be sure.

The lower right cell represents the situation where the company has an experience base with breakthrough innovation and has developed a highly functioning BI management system, but the company's capacity is constrained. The key in this situation is to power down, but not extinguish, the BI capability. It is important for the orchestrator not to lose that experience base by allowing the group to disband, suspending all of the projects, or cutting all funding for the effort. That's what companies have done in the past, and they tell us they have wasted far too much time, talent, and money.

Depending on the sources of the constraint, the orchestrator faces different options. If it's a situation of lack of willingness (the new senior leader does not wish to compete on innovation), the orchestrator can try to influence that through discussions, exposing the leadership to experts in the area, and making a case for the consequences of ignoring innovation. This is the key reason that the orchestrator should keep a rather elaborate set of metrics that track many dimensions of the BI portfolio's performance over time. Then she can be ready to answer the "What have you done for me lately?" question, which is asked every time there's a change in leadership. (At times, the BI portfolio may be performing well on some of these but perhaps not on others.) The options investment mentality is critical, so that money is spent only as a glimmer of promise emerges from any given project. The orchestrator can also revisit the BI group's mandate to focus on aligned opportunities for areas of the company that are receptive. Some business units may be experiencing healthy growth, and others may be suffering and may want help to fuel growth.

If instead the organization is facing profitability challenges or a lawsuit or a potential acquisition that is distracting a committed leadership's attention and resources, the BI function can continue with its portfolio, but pace it more slowly or reduce its size, pending the resolution of these external threats. This was surely the case at Corning when its stock hit $1.10 on October 8, 2002, but the CEO, Jamie Houghton, proclaimed that innovation was the key to Corning's future success. The company didn't have much to invest in innovation, so they paced it, but they did invest.

Finally, the lower left quadrant of Table 7.2 offers ideas for those wishing to initiate the development of a BI capability but face both constrained capacity and a lack of experience in the DNA functions. Typically these initiatives come from middle management ranks.[4] These people are emergent orchestrators, sounding the call for the need for innovation in the company and volunteering to make it happen. They may learn about innovation from books they read, conferences they attend, or colleagues they meet in other companies through professional organizations.

Their first approach is to try to educate others senior to them in the company about the need for innovation. They may bring in

outside experts to speak or direct their leadership to specific read-
ings or Web sites. They may also use a small amount of money
to get several projects started and proclaim success based on any
amount of progress that is made. In some cases, this is a fruitless
effort. When they're just apathetic, there's room for the orchestrator to work. But when senior management simply cannot
be convinced and they are against innovation (and some absolutely are), it's not worth the effort. There are better uses for the
orchestrator's energy and plenty of other places that value that
rare talent. Indeed, in one company, the president publicly stated,
"No investment in innovation; we've got to control costs." However, one or two middle managers (orchestrators) realized that
once costs were under control, the president would want to know
where the innovation was. So they continued to work on innovation, using discretionary resources and shared people to keep some
momentum going.

Challenges of Pioneering Orchestration

Clearly the orchestrator's role is crucial to any company's ability to sustain an effective innovation capability. In many ways, it
is an exciting, rewarding role, but it has its dark side. Two major
challenges emerged from our interviews with orchestrators. We're
certain there are others, but we highlight these because of the personal tolls they took. Orchestrators who are pioneering the development of an innovation function in their companies may find
these familiar. But once the innovation function is recognized and
on course, these problems should dissipate.

Working with Innovators

Many times, people who gravitate toward breakthrough innovation work can be difficult to work with (more on that in the next
chapter) due to their independent, maverick orientation. While
the orchestrator is expending lots of energy managing upward (to
senior corporate leadership) and outward (to business unit leadership)

to ensure that the innovation group's work is aligning with and influencing the company's strategic intent, those who work for the orchestrator, or the project leaders who comprise the portfolio, may not care about strategic intent at all. In fact, they're frequently rebellious sorts. Additional energy is required of the orchestrator and the DNA leaders in general to manage these people. They're important, but they've got to be aware that they're valuable to the company only insofar as they work with, and not against, the shared objectives about long-run breakthrough innovation that company leaders hold. We find that incentive forms of compensation are less valued by these individuals than is receiving credit for their contributions to the company. They want to be recognized.

In addition, these people tend to balk at any sort of processes installed in the innovation function. They've battled operational excellence-based processes their entire careers. "We need guidance, not process," one group told us. The orchestrator and the DNA leadership team have to develop appropriate processes, like the learning plan we described in Chapter Four, and ensure that the innovation system is not bogged down in processes for processes' sake. Innovators have an innate sense of impending bureaucracy. For them, innovation systems and processes speak of hierarchy and bureaucracy, which means they are against it. It is important, then, to demonstrate and otherwise convince these mavericks that the processes will assist and support them, not make their lives more difficult.

The problems of managing the innovation staff can be brutal. In one of our sample companies, a new orchestrator was brought in to take over from the person who had started the innovation initiative. The two leaders' styles were dramatically different. The first spoke in visionary terms and focused heavily on developing a collaborative culture, and left everything else to the teams. The second, having come from an operational group where he was quite successful, instituted a system that he was more familiar with: formality and rules. While this second orchestrator worked hard to raise the visibility of the group within the company, the style differences were too difficult for the group to cope with; they preferred an open, interactive, and impromptu style that fit their personalities. In addition, the second leader's external focus was a dramatic difference to the innovation staff, and they did not like the lack of personal attention. The second leader was clearly a mismatch for the strong, self-directed innovation staff. After two years, the second leader was fired in a mutiny-type setting. The innovation staff complained to the CTO about the leader's style. That was it.

The Emotional Roller-Coaster of Working with the DNA System

Commitment from the top can seem on-again, off-again. People from outside the innovation group lack patience and want to see results. People within the innovation group may become demoralized as failures mount, or they may lack patience as the path forward on a number of the portfolio projects continues to be murky. The orchestrator is part of the effort to move the organization from an invention culture (where progress is measured in terms of the number of patents obtained) to an innovation culture (which requires incubation, acceleration, and all the rest). Without a strong management system in place and without organizational support, orchestrators can get burned out. It's critical for orchestrators to assess their organization's capacity and be realistic about what they can accomplish. Sometimes they can, like Don Quixote, pursue a vision of a highly functioning innovation management system when the company is simply unable to accept it. The conversation that follows is taken from an exit interview we held with one of our orchestrators who decided to step down after a lengthy period of time battling a senior leader who was focused on short-term cost reduction as the only way to compete:

> *Question:* Can you talk to us a little about your reason for not taking the new leadership role?

D: First, I'm worn out. This is a role that has emotional ups
and downs. My view is that you shouldn't have this job for
more than three years, and I've had it for four and a half.
You need fresh people coming in, and I started losing my
enthusiasm a year ago. I was fighting against barriers. More
important, I didn't see the top management support we
needed to really make [breakthrough innovation] a success.
I know this doesn't hold with everyone's theory about what
it takes to run a BI group, but it's a lot of work when you
don't have that support. But [the new CTO] is going to be
much more supportive. We also have some changes on the
executive council that makes it more technically oriented,
so this could be positive. Nonetheless, it's time for me to
move on. If I want my career to advance, I can't stay where
I'm at. Here are some thoughts about what I'd do differently if I were starting out again, knowing what
I know now:

When I started this job, we were in the middle of the Internet
bubble, and innovation was all the rage. There was a mind-
set that permeated the country about being a revolutionary
and being different. This sold quite well in the United States,
but it fell flat in Europe. It took a while to develop some
good relationships with our colleagues in our European lab,
who we were trying to encourage to become risk-oriented.
They are lower key and are more technically focused. Next
time I would give more attention to the cultural differences,
because they turned out to be a key barrier to our success.

I would have gotten a lot more clarity earlier on the strategy for
[my group] from my boss, the CTO. I would have worked to
get public endorsements. I would have worked on this much
earlier. We actually had this finally happen in the summer
2003. His ambivalence was sending mixed signals to the
organization.

Question: Now that all seems so clear and obvious, but is it possible that during your first two years, you were so excited

that you didn't want to know? What would have happened if [the CTO] didn't provide the words you wanted?

D: I don't know. Here's a bit of a confession. I made the assumption that [the CTO] and the top leadership agreed to have a [radical innovation] group and they would want the group to actually work on radical innovation!! Signals came out over time that caused me to realize that [the CTO] had very different views of what innovation is or what types of innovation we should be engaged in.

This was a big revelation for me. Additional signs indicated that we were not taking enough of a low-key approach. You're pushing the boundaries a bit too much. You need to tame it down a bit.

Question: How much of it for you was, "If I make progress and show results, [the CTO] will come around"?

D: A lot of it.

Question: If [the CTO] sat you down and told you [the company] was more about incremental innovation, would you have quit or plowed forward?

D: I would have plowed forward but have been more politically astute. I would have taken more of a skunk-works approach. I would have kept it quieter. I wouldn't have raised the visibility of [our group] as much as I did. I would have sought out help. I would have thought through the political issues.

Very related to this is that I've thought often about my role in educating [the CTO] and others about innovation. The first question is to understand why there wasn't as much support for it and getting a fair understanding of what's behind that. If I then still believed it was the right thing to do, I would have engaged in more education. Not everyone understands innovation and growth in a company. [Our CTO] had a misperception that innovation is associated with chaos. He felt that if he allowed the organization a bit too much freedom to innovate, you would lose your focus and lack operational effectiveness.

One of the things that I would have liked to have done better is to communicate more, and more effectively, to all of our stakeholders. I didn't realize that when you're trying to drive change, how difficult it is to communicate the message. I only began to appreciate this challenge in the last two years. I thought it was pretty straightforward, but it's not. People have to hear the message five different times using five different media.

Conclusion

D obviously learned from his experience, and we all can too. The orchestrator is crucial to the well-being of an innovation function. Sometimes, in smaller and medium-sized companies, the orchestrator's role can be played by a senior person. In larger firms, the orchestrator assists the innovation senior leader by monitoring and diagnosing the innovation system, and, in particular, helping to manage the moving boundary between the innovation and operations management systems. We firmly believe that this role has failed to be recognized in the past, and by surfacing it and describing it, companies can begin to recognize the need for it and put an orchestration role in place.

Next we turn to the real practicalities of how a company initiates an innovation function. We find that there are predictable patterns companies face as they attempt to put this in place, and we again invite you to learn from others' experiences so you can avoid missteps and build a stronger innovation function.

Questions for You

1. What is an ideal orchestrator for your organization? Be specific in terms of background, experiences, skills, and personality characteristics.

2. Now identify the person or persons currently serving in that role (it may be you!). How closely does he or she fit the requirements you listed in question 1?

3. What are the pros and cons of having a single orchestrator versus a team of them?

4. What clearly does the orchestrator in your company align the DNA activity with the company's current capacity? How willing and able is the orchestrator to flex the DNA activity as capacity changes?

8

GETTING STARTED

Initiating and Maturing an Innovation Management System

You may be asking yourself, "Where do we start?" It might seem, at this point in the book, like a complex, burdensome, and expensive set of tasks. After all, we've got three major building blocks, each with its own structures, processes, governance mechanisms, funding models, and on and on. Then you have to knit it all together at the DNA system level and add that invaluable orchestrator to work all of the interfaces. You might feel that your company couldn't possibly make such an enormous commitment, and even if you decided not to undertake the whole system at the outset, the question still is, "Where do we begin?"

It may seem overly burdensome to you now, but keep in mind that we are not talking about undermining your current systems or processes, or about changing anything regarding your current systems for maintaining and improving your ongoing businesses. We are talking here about adding a much-needed capability—a capability for major breakthrough sorts of innovation to the activities of your firm. This does not mean replacing current systems for managing ongoing operations or product improvements. The problem that established firms face is that they've not focused on developing a capability for breakthrough innovation. Our concern revolves around leveraging that rather small investment into substantial new businesses.

Let's put this in perspective. R&D as a percentage of sales for most companies is less than 4 percent. Some biotech or high-tech companies, it is true, may invest as much as 15 percent or more, but they're the exception. The question is, How do you leverage that small investment into fully developed new businesses? It's a small investment for ensuring your company is around for a long time. The real costs are incurred in the effort required to develop an innovation mind-set, an innovation-supportive culture, and an infrastructure for allowing the DNA system to operate. But that's where the benefits are realized as well.

Many authors claim that established companies are unable to successfully commercialize breakthroughs,[1] but the fact is that companies have not dutifully tried. Mature companies have relied on luck, timing, and the talents and perseverance of very rare individuals (mavericks, champions) to make breakthroughs happen. This results in one-off, infrequent occurrences that do not leverage the firm's cumulative knowledge, capabilities, or management prowess. Companies can do better. In our view, breakthrough innovation is the next major management capability large companies will claim as their priority, much like the quality effort was claimed in the 1980s. We now know that investing in breakthroughs, even with their high levels of risk, yields positive financial results over time.[2] So far in this book, we've presented an elaborate description of a breakthrough innovation capability in terms of all of the elements of its management system. The challenge ahead is to figure out how to implement it in your organization. The road to success begins with a single step.

One of the biggest surprises of our study is this: the challenges that companies that are building a breakthrough innovation capability face are rather predictable over time. There are specific issues that emerge when you're just getting started. Within a year or two of your effort, new challenges arise. As your efforts evolve toward a sustainable system, a different set emerges. Over the course of our eleven-year research program, we've studied many of these companies. It is amazing to note the patterns of growth and challenges that companies face as they work to build breakthrough innovation capabilities. If you could learn from their experiences, your company may be able to sidestep some of those heartaches.

That's the purpose of this chapter: to make you aware of what your company may face and help you leverage others' learning. We want to help you plan your path forward.

We note (actually, we sort of imposed) four distinct phases of development of a breakthrough innovation capability, and in sharing those here, we hope to help you move through the initial ones more smoothly than some of those who've blazed the trail. Then you can get your company to that mature breakthrough innovation capability level with less time, money, and aggravation.

The Phases of Developing a Breakthrough Innovation Capability

For purposes of helping you manage through this process, we identify four eras in a company's effort to develop a breakthrough innovation capability. The first is *setting the stage*. That's the time when the company recognizes the need for innovation. There may be a critical triggering event, like a noticeably large missed opportunity (IBM's situation), a dramatic drop-off in current business success (Corning's situation), or a change in the leadership of the company (as at 3M and GE in recent years). Or maybe instead someone from middle management wonders why the company isn't more focused on innovation and has a passion for making it happen (as at MeadWestvaco, Air Products, Shell Chemicals, and Albany International).

The next phase is *getting started*. Someone has to start building a capability. What are the starting conditions? What's the best way to begin? What are all the elements of a management system that have to be considered given that no one can address all of them immediately? What options do you have?

A year or two after companies get started, we note that BI initiatives begin to face specific sorts of challenges. So we note that *maturing* is a third phase in this process. Given that a number of the same challenges arose across many of our companies, we think you should be aware of them and understand how companies that ultimately continued to build a mature innovation capability addressed them.

We'll describe what we've learned about each of these phases in some detail in this chapter to help you understand how to get started or, if you already have been at it awhile, to help you resonate with other companies' experiences and get to maturity as quickly as possible. Finally, we'll consider what a *sustainable*, mature BI capability is. What does it look like? How do you know when you've gotten there? How do you maintain it? The risks to the system is another consideration, but that's left for Chapter Nine.

Setting the Stage: Why Undertake to Build a Breakthrough Innovation Capability?

When we knew we wanted to begin this study, our first order of business was finding companies to study: those that met the criterion we needed: a declared strategic intent to build a breakthrough innovation capability. We interviewed a number of companies. Many well-known large firms did not qualify. Why do some undertake developing such a capability and others not? Not all companies want to compete on the basis of innovation, and not all companies can. Some want to and do not have what it takes.

With a set of companies with a declared strategic intent, we went back and asked them how that strategic intent was developed, who was responsible, why it developed, and how the strategic intent was initiated. So if you are considering adding a breakthrough innovation capability to your arsenal, how should you prepare? What are the considerations? How should it be scoped? What has prevented you from doing it before? These are the issues that you must address prior to getting started. Here are the reasons our companies gave for initiating a breakthrough innovation program:

Growth Strategy

"Paid down debt, squeezed costs. Now what?"

"Current growth rate too low for our industry."

"Our focus on incremental innovation keeps us in a competitive pinch."

"Acquisition strategy for growth is not producing returns."

"Need to create new white space businesses."

"Need a holistic growth strategy incorporating organic growth and acquisitions."

Financial Return

"Need to increase the price-to-earnings ratio."

"Stock price has flattened; need to do something."

"Need a better return from R&D investment."

Technology Strategy

"Make technology the key driver of competitive strategy."

"Build technological capabilities in a few focused areas for major future impact."

"We need fewer projects and bigger hits."

Competency Development

"We need a long-run competence for growth."

"We want to create business value, transform business models, build core technologies."

"We want to learn how to link technologies and business models."

"We want to build entrepreneurial behavior into our organization."

"We have strayed from our breakthrough innovation roots and need to return to them."

Diversification

"We need 20 percent of our portfolio in white space innovation."

"Need to leverage our core competencies into new markets."

"Need to leverage innovations from one part of the
organization to another."

Defensive Moves (Internal and External)

"Our current businesses are dying off."

"Why have we missed the last few big market opportunities?
What's wrong with us?"

"External market shocks have threatened our organization's
health."

"We need to legitimize R&D's role in the organization."

"Need to placate R&D given recent budget cuts, so start this
new program."

Do they all make sense to you? They didn't to us!

There were similarities as well as differences in the reasons companies gave for wanting to develop or improve their breakthrough
innovation capabilities. Nearly all of them reported growth objectives. The focus on cost containment of the 1980s and on growth
through acquisition has given way to a period in which companies are seeking growth through organic, or internally generated,
means. Most of them also described their desire to escape the "lots
of small stuff" trap. Competitive intensity and low margins were
driving these companies to find new territories and new ways to
bring real value to the market. A final commonly held theme we
heard came from the R&D organizations, which had been reeling
from being slaves to business units' immediate needs and funding
coffers. R&D leaders expressed the desire to work further into the
future. Morale was low, and the loss of R&D talent due to lack of
interesting projects was a concern for some of these companies.

Companies differed on a number of triggering issues as well, and
these are the sources of confusion. Some were seeking fairly immediate financial returns from their breakthrough innovation initiatives,
while others expressed a need to develop new competencies that
would yield financial rewards over time. Some saw this as a new way

to view R&D management, while others adopted a more holistic view that we might call a *new business creation perspective*. In other words, they recognized the importance of R&D and invention, but they also realized that invention was just part of the picture. (The latter companies are the ones that developed incubation and acceleration capabilities the most rapidly, as you might imagine.)

Some began the innovation-building journey based on a defensive posture (for example, "How'd we miss that last wave of opportunity?"), and others were more offensively oriented (for example, "We view these growth platforms as the wave of the future, and we expect to be leaders in those areas, so we're investing now"). Finally, we noted big differences in the sense of urgency that accompanied BI initiatives. In some cases, the companies were in dire financial straits, and panic had to be managed. These companies had to be concerned that the expectations they set for the BI initiative were realistic. In other situations, the urgency was too low. We saw lots of experiments but not enough impatience on the part of leadership to focus on learning from mistakes, redirecting in a systematic way, and improving the output as a result. So in several cases there was lots of flailing around, but no results and no pipeline at the end of the four years. There was a continuously evolving infrastructure that was really one experiment after the other accompanied by wide variations in budgets over the years, or no real budget at all.

Whatever the reasons, for the BI capability to become embedded as an organizational system, there needs to be motivation or a triggering event. IBM's Lou Gerstner issued a call in 1999 when Big Blue once again missed a major opportunity: "Why do we consistently miss the emergence of new industries?" he asked. That question, from that senior-level leader, initiated a self-analysis that ultimately was the basis for launching the emerging business opportunities (EBO) system at IBM.

The felt need, the urgency, doesn't have to originate at the top of the organization. At the beginning, senior leadership may in fact be one of the company's major constraints to developing an innovation capability. The initiative can start with middle management developing a localized innovation hub. It's true that unless senior leaders become convinced over time, the innovation function can never take root and never flourish. Middle management can plant the seed and tend to it. But it'll be perceived as a weed unless and until senior leadership comes around to supporting it. Senior management can learn from middle management's initiatives. In fact, in thirteen of our twenty-one phase II companies, the initiative started from within the middle management ranks. Nevertheless, innovation as a full-fledged function, with its own management system embedded in the mainstream company, cannot take root without senior leadership support and attention.

Whatever the reason, the need for developing the BI capability has to be felt by everyone connected to this new initiative, and realistically, the need for the capability needs to be felt by the entire organization. Otherwise there won't be any urgency, and any other initiative will capture organizational and management attention, taking the air out of the innovation balloon.

Activities Associated with Setting the Stage

How do you successfully navigate your firm to decide whether to build a breakthrough innovation capability? The purpose of setting the stage is to undertake a self-analysis before getting started. This involves:

- Motivating the need for breakthrough innovation by stating the problem
- Clarifying the mandate, scope, and objectives for the breakthrough innovation function
- Conducting a capacity audit
- Communicating it broadly

Motivating the Need for a Breakthrough Innovation Function: Stating the Problem. First, identify the drivers for investing in the development of an innovation function. Why are we doing this, and what are we expecting from it? Drivers, or reasons for investing in developing a breakthrough innovation capability, could be any or all of the following:

> "We could be doing better at creating new businesses."

> "We miss emerging business opportunities [or missed another technological discontinuity]."

> "We've atrophied in our innovation ability, and others are passing us by."

> "We're too mired in the urgent, but we're missing the important."

> "Our employees want to innovate and use their skills. It'll help us retain the best talent."

> "We need to break out of the commodity markets we've gotten ourselves into."

> "We cannot continue to acquire (federal regulators are telling us). So if we want growth, it has to be organic growth."

> "We've saved and saved. Now we need growth, big growth."

The drivers set the tone, the degree of importance of the innovation function. Make sure they make sense and are worthy of the investment.

Clarifying the Mandate, Scope, and Objectives for the Innovation Function. What do you expect this group to achieve? Will it be responsible for white space businesses? For businesses five years from now? For finding any new growth opportunity, regardless of timing horizon or alignment with current operations? What are the responsibilities of this group relative to other innovation groups?

Realistic, appropriate objectives for breakthrough innovation are not about immediate delivery of financial impact to the bottom line. They're about creating new platforms for growth over a substantial time frame, with an eye toward future financial reward. The reality is that breakthroughs aren't easy, and they take time. To initiate a capability from scratch and expect financial results in six to twelve months is a setup for failure.

Your objectives should be based on your company's basis of competitive advantage. Does it win on the basis of process innovation? Business model innovation? Product innovation? Are you a materials company such that you could leverage an innovation into many applications? Or do you succeed on the basis of combining many diverse technologies into a system-level solution?

We observed a lack of alignment of objectives between senior and middle management. In several cases, senior leadership is looking for financial returns, and middle management is looking for new creative outlets or ways to make a bigger difference in the world. Or middle management wants to undertake wholly new business opportunities, and senior leadership believes all growth should fall within the current lines of business. Most senior managers are not business builders; they're business maintainers. The structure of senior leadership's compensation package arose repeatedly in our workshops and think-tank sessions and was identified as a potential showstopper. Since senior leadership turns over fairly frequently and has incentive compensation tied to current or near-term earnings, it appears nearly impossible to believe that breakthrough innovation investments can be made systematically.

The problem with a lack of alignment of objectives, between senior management and those who will execute the BI mandate or across the company, is that performance measures that are established will not match the reality of the role. Again, this is a setup for failure before the first dollar is invested.

Remember the orchestrator, D, in Chapter Seven who realized that his senior leadership understood breakthrough innovation to mean a very different thing than the orchestrator himself did, and after three years? These issues could (and should) have been clarified earlier. It's at this stage, before any real investment is made on the part of the company or its employees, that these difficult discussions have to occur.

Any innovation capability needs multiple objectives. One may be to win in the marketplace, but another may be culture change internally, and a third may be the development of new competencies or entrée into new market domains. There are so many setbacks that you may become demoralized on any given objective at a point in time. With multiple objectives, you can show and report progress along many dimensions.

Chosen objectives should have both short- and long-term components. For example, a short-term objective might be number of ideas in a breakthrough innovation capability pipeline or number of ideas that result in BI projects. Longer-term objectives typically have some sort of financial measures such as revenue growth or increasing margins. As we have been discussing throughout the book, unless measures are chosen, agreed to, and committed to, mandates and objectives will change as business pressures impinge on innovation activities.

Conducting a Capacity Audit.   Before undertaking an investment in developing a new capability, it's important to assess whether it's realistic to think your company can accomplish it. Has your company always competed in terms of being a fast follower but never a market pioneer? Are you willing to commit to hiring the technical and entrepreneurial talent you'll need? Do people need to be reassigned? Does your company reward risk taking? Or is there a fear of failure and its consequences in your organization? If these are characteristics of your company, you will likely not get the participation you hope for when you post new job openings in innovation. Is your organization under so much stress (everyone has two or three jobs, there's a high level of management turnover, or the market for your products is extremely volatile) as to make it impossible to concentrate on developing new capabilities? What are the competing priorities, and how likely is it that they'll require more attention than this one? This doesn't mean it's impossible, but it does mean that you have to consider the additional constraints your organization is facing and start small. You are operating under a constrained capacity situation.

Second, consider why you haven't been able to do this before. What are the causes of previously failed attempts? In one company we studied, the CEO noted that the company had just missed a major opportunity, and he could not understand why. He set up a task team to conduct a root cause analysis to figure out how this had happened and report back to him. The team conducted interviews and analyzed the data for several months. One of their most striking findings was that senior leadership said they wanted BI, but when the task force looked at the senior leaders' calendars, almost none of their time was spent on innovation for the future. In fact, almost none of their time was spent on planning, or even imagining, for time horizons more than a year away. This sent a much clearer message than their words of support for breakthrough innovation. Many other root causes were identified, including these:

- The management system rewarded execution directed at short-term results and did not place enough value on strategic business building.
- Leaders were occupied with their current markets and existing offerings.
- The business model emphasized sustained profit and earnings per share improvement rather than actions to drive higher price-to-earnings ratios over time.
- Leaders' approaches to learning about markets and using market insights were inadequate for embryonic markets.
- Leaders lacked established disciplines for selecting, experimenting, funding, and terminating new-growth businesses.
- Once selected, many fledgling ventures failed in execution.

Sound familiar? If you were to conduct a root cause analysis, would your learning be similar?

This analysis helped clarify to the team and the CEO what actions and resources would be required if they truly wanted to build a breakthrough capability. Are your objectives consistent with building a BI capability? Are they of a longer-term and high-magnitude orientation? Do they recognize and build in uncertainty? Do they recognize that failure is not only an option but likely a reality within your portfolio? Can the resources you provide be consistent with this set of objectives? For example, start with a promise of five years of funding rather than one. You need a message of consistent commitment so the group can start doing the right things at the outset and not worry about getting that first project out the door.

Hence, it's important to have a realistic picture of your company's risk tolerance, investment propensities for new projects, cultural fit with innovation, and, especially, your company's punishment or reward system for failure since BI naturally entails some amount of failure. This kind of assessment will indicate the degree to which the innovation initiative should be scoped. One reason for the early failure of such initiatives is an overly optimistic scope for the BI system developed in an organizational environment antithetical to it. The questions at the end of Chapter Two should help you start your capacity audit.

Communicating It Broadly.   Finally, the objectives for the BI group should be communicated broadly across the organization, and a plan should be in place for ongoing communications. If there isn't understanding of the initiative, why the group is in place, and what its role is supposed to be, especially in relation to other innovation-related groups in the company, there'll be plenty of misinterpretation and resistance. BI groups in our companies were getting calls for help on issues that were not in their purview and requests for funding projects that weren't breakthrough opportunities, or they were viewed as threats to other parts of the company. When the latter occurs, jealousies, stinginess with resources, and even sabotage can take place.

Moving On from the Groundwork Laid

This initial planning phase should lay the groundwork for a healthy start. Objectives are clear, realistic (and exciting), commonly held, and well communicated. Now it is time to move into action. We realize that things don't happen so rationally in many, even most, organizations. That's okay. We're stating the ideal approach here. As long as you're aware of the issues that should be addressed, you can watch for them as the breakthrough innovation initiative unfolds in your own organization.

Getting Started

Begin with understanding that what you are developing is a capability, not a process. A capability means that you've got to put the entire management system in place; a process is a sequence of steps to accomplish something. Although you cannot get it all in place at once, it's helpful to know what the end game is at the outset, so you can decide how to proceed rather than getting caught by surprise. These aspects of initiation surfaced as important in our study:

- Build a culture for innovation.
- Find people: project leaders and support staff, including coaches.
- Set up the infrastructure. What programs will you run? Where will you be located? Who will you report to?
- Find, co-opt, or generate the first few projects to start working on. Focus on process development after you've accumulated some experience.
- Announce your existence.
- Explore resource models.

Let's look at each of these a bit more carefully.

Build the Culture for Innovation

We've made the point repeatedly that the organization's innovation subculture will differ from the dominant culture that most organizations experience: a culture of operational excellence. Operational excellence requires strict adherence to plans, budgets and schedules, customer intimacy, and leveraging what is known. The innovation subculture operates in a world of high ambiguity and has to bring some level of discipline to chaos. Table 8.1 provides a contrast of the two worlds.

To be accepted, the BI group has to educate everyone in the company about its activities (senior leaders, rising leaders, potential project team members, and potential idea contributors, who might be anyone in the company), what a culture of innovation is like, and what behaviors reinforce that culture. For many companies, this is foreign and unnerving, which limits contributions to the BI effort. As one of our participating company members running the BI capability system told us, the recent focus within R&D on servicing near-term business unit requirements had meant that even in R&D, people were no longer used to a culture of innovation: "We need practice; we've atrophied." In several of our firms, outside consultants, experts on innovation management, were brought in to help expose all members of general management to the world of innovation: ambiguity and how to navigate it. Others approached this by assigning senior leaders to sit on advisory boards of some of the early projects. "They've got to learn what this is about," one very senior leader running the BI initiative told us. "The best way I know how to educate them is to get them involved in these businesses as board members and let them interact with the team and with me regularly."

Members of the BI groups in some of the companies began running idea generation workshops to get people interested and involved in innovation again. Much of the time spent in those early workshops, we're told, was not directed to generating ideas so much as to talking about innovation and why it was important for

Table 8.1 Motivating Reasons for Initiating a Breakthrough Innovation Capability

Management Issues	Operational Excellence	New Business Creation
Strategic drivers	Financial and market performance.	Strategic intent and opportunism.
Planning horizon	Quarterly to two years.	Three years and beyond.
Type of Innovation	New products to extend existing businesses and gain or maintain competitive superiority in current markets.	Platform and breakthrough for growth and renewal to create new market spaces.
Culture (behavioral and attitudinal norms)	Clear rules and processes; clear norms on power and division of responsibilities; task oriented. Variance and variety-reducing activities. Actions based on current knowledge.	Few rules, outcomes oriented. Control based on vision, goals, and values. Cultivation of the novel; experimentation. Variance and variety-enhancing activities. Competence-destroying behaviors in order to find new competencies.
Risk profile	Risk averse with focus on system efficiency.	Risk mitigation through staged learning and investment.
Opportunity selection	Customer driven using market research tools.	Vision and possibilities tied to strategic intent
Investment timing, revenue focus, and success criteria	New products in six months, profit-and-loss management with in-year revenue streams. Earnings per share, share price growth.	New businesses in three to five or more years. New business platforms, new customers, top-line growth and return on investment over time. Portfolio management to hedge bets.

the company. They established definitions and a common vocabulary regarding innovation (aligned versus white space, incremental versus breakthrough, game changers, or step out or horizon 3) so that everyone in the organization could begin to operate from a common understanding of these different worlds.

Several of our participating companies required all rising leaders in the organization to spend some time working in the innovation group, just as they had rotated through marketing, managed a product group, worked in operations, and gained experience in many other assignments in the organization to prepare for general manager roles—an encouraging sign. It meant these companies were finding a way to ensure that all leadership would have experienced both operational and innovation worlds and so would be better prepared to understand the innovation world as a legitimate complement to the world of operational excellence.

All of that said, here's a note of caution: company cultures differ. Operational excellence is generally dominant, but there are many ways it can play out. The norms of behavior in one company will not be accepted in the next. This means that any attempt to build a BI capability in a company must respect its culture. You cannot impose an innovation system on a company that does not respect the way things get done. Here are some illustrative quotes from our participants. Note the contrasts among them:

"Our culture is to buy and build, not create and build" versus "We will invent our way out of this financial slump."

"We're process oriented here" versus "We're relationship oriented."

"We are not a democracy here" versus "No one has to do what our CTO says. They just ignore him."

"If you overpromise and underdeliver, you're dead" versus "We all know we're learning here."

"If you take a risk and you're wrong, expect a pink slip" versus "If you don't take risks here, you cannot move up."

Clearly, if you are in a buy-and-build company, as noted in the first quote, you would not propose to build an internal R&D capability requiring hundreds of millions of dollars. You'd build your BI group to ensure adequate resources devoted to exploring external sources of technology and other opportunities. But you surely would not set that up as the main use of discovery resources if you were in the second company quoted. You have to work within your company-specific culture and introduce organizational members as appropriate to an innovation mind-set.

Find the Right People

What issues do BI leaders face as they start to staff their innovation function and project teams?

Innovators Who Can Be Difficult. BI team leaders and those who initiate ideas can be a demanding sort. We've seen disgruntled, rebellious, underused, curious, creative rule breakers gravitate toward the innovative organization because they can leverage their creative energy into leading-edge projects. They do not conform to preset processes and are hypercritical of many of the norms and the people in the organization. Egos abound. Sometimes they border on cynical, and their criticality can be discouraging. They are high-maintenance individuals, even prima-donna like, and can sap the energy of the BI group leader. They can even be alienating. Nevertheless, the organization rewards and encourages these people because they tend to solve problems and generate innovative ideas.

Although they're typically brilliant, their divisiveness and individualistic orientation makes them difficult to manage. When innovation is accomplished by relying on strong, persistent mavericks working against the organization, the organization not only accommodates but can encourage them. However, when you are building a group or a new function, its members must behave like part of a team and have their company's future health at the top of their agendas. When these people are included in the conversation regarding the company's long-term strategy and their group's role in helping execute on that strategy, they can be converted from detractors to advocates. We find that many times, BI group leaders fail to include them in these conversations simply because they are used to treating them as independent mavericks.

Companies don't really know how to handle these kinds of people, but they keep them around because they tend to be problem solvers and innovative thinkers. However, their cynicism can be discouraging to others. Those who have high entrepreneurial propensities probably will not stay with the company for very long in operational excellence–related roles, although they are well suited to BI–related roles. The challenge is to identify those people, encourage them to work in BI activities, and then reward and develop them for careers in innovation. We have seen seeming malcontents blossom when put into innovation roles. The challenge for human resources is to develop a selection, reward, and retention program suited to them, different from career programs associated with operational excellence roles.

Gary Einhaus at Kodak's Systems Concept Center, Kodak's BI activity center, was able to forge a high-performing team environment with innovators, but it operated in quite a different way from the mainstream organization. The SCC staff personnel were highly talented and motivated. Many members could be described as difficult personalities who harbored much discontent with the current state of the company as it pertained to the business and product trajectories. Moreover, the group was querulous and intransigent by nature. In particular, they consistently fought any imposition of business process on their creative efforts.

The SCC was led for the first six years of its existence by Gary Einhaus. The method of work and review had an element of performance art. Those with master of fine arts degrees, ethnographers, and design artists skilled in the presentation of novel information were highly advantaged in the early stages of an entrepreneurial project. A large part of the efforts of those who worked in the group were oriented toward ideation and bringing new stimuli into the corporation.

There were several business research professionals in the group who took on roles of "ideation event facilitators." These individuals had many years at Kodak and were adept at capturing the results from ideation sessions. The structure of the sessions varied, but normally the participants were individuals from the SCC's alpha team (the discovery group) and invitees from R&D and the business units. The alpha team was a loose configuration of rampantly curious individuals who moved in and out of the innovation center. The core of that group was a closely knit assemblage of individuals with M.F.A.s, M.B.A.s, and Ph.D.s in science with noteworthy backgrounds at Kodak. Their specific charter was to bring innovative stimuli into the greater sixty-member group and initiate new and creative offerings. These sessions were useful in connecting outside information with business and science information generated internally. The alpha team worked closely together and informally fed their results to the greater firm through "trends meetings." Although the individual team members were a diverse and contentious group, they were united in their commitment to the vision and mission of the SCC, forming the basis of a highly effective team, albeit looking different from one that might be associated with operational excellence.

Shortage of Skills in New Business Creation. One critical skill set is new business creation. We note a severe shortage of such expertise in organizations and in the marketplace generally. Many have told us that their dominant culture drives these people away (and likely serves to sour the individuals described in the previous section). A person who knows how to build new businesses and prefers a large organization to the start-up environment is indeed rare. These folks have to be entrepreneurial but thrive within an organization. They must be able to negotiate, network, barter, and articulate their value to the company over and over.

At initiation, most of our BI group leaders did not have a problem finding people to help. Many people interested in innovation exist in most organizations working in jobs they don't like. They come out of the woodwork to join the innovation group when it is announced. Indeed, not one of our BI group leaders complained about the lack of internal applicants for job postings in these groups. So the desire is there, but few have experience at the outset. Accessing people externally and tapping those who have built businesses in other parts of the organization are approaches most of our companies took. Otherwise the team tends to learn together slowly, through trial and error.

Lack of Innovation Career Path.   One of the reasons talent in new business creation is so rare, we believe, is that there is no career path in this area. One of our company representatives relayed the story of Mitch, a Harvard M.B.A. who'd joined the firm at the same time as several of his classmates. Mitch had an entrepreneurial flair, and within a year or two of joining the firm, he was tapped to lead a promising but young and risky new business venture. He declined initially, indicating to his bosses that his colleagues from Harvard who were in the operating units were more likely to be promoted given the high likelihood that they'd meet their target growth objectives, whereas Mitch could see no certain path to profit, only promising ground to till. The leadership was stunned and set about promising Mitch security and potential for promotion, especially based on the extra exposure he'd have to senior leaders as he worked to develop the new business. Mitch took the offer, did not succeed, and didn't rise through the ranks as quickly as his Harvard colleagues did. So the story goes. It's difficult to find the right talent. When you do, you'd better take care of it.

Underdeveloped Breakthrough Innovation Leaders.   Most BI group or portfolio leaders in our study expressed that they felt they were undertrained for the job. Only six of the twelve had any experience building businesses themselves in the company or another organization. The rest believed the company needed this capability and initiated it or stepped up to the assignment, but they had no clue what to do. Most came from the R&D organization. We expect that as time goes on and more firms build groups for new business creation (as they are), there'll be more experienced talent available. But right now, it's in very short supply.

Garnering people who are motivated and able to contribute is key to jump-starting any new initiative. In the case of innovation, established companies are floundering. Many innovative types have fled the large company environment due to the dominance of the operational excellence management system. They cannot stand it. So finding them and protecting them while you are getting your innovation function set up is of key importance.

Set Up the Infrastructure

Some of the companies in our study did not have an identified organizational group responsible for the breakthrough innovation agenda. "It's everyone's responsibility," we heard. "It's part of our culture," was another reason they mentioned. The idea was that if senior leadership proclaims innovation is important in the company, it'll happen. As it turns out, that's not so true. Companies that claimed innovation was part of their culture were absolutely right: we observed lots of innovation. However, not much of it was breakthrough innovation. To commit to developing a BI capability, a company needs mechanisms for tying the innovation investments to the company's strategic intent, populating project teams with the kinds of skills needed, overseeing a project's progress in relation to other projects in the portfolio, experimenting with many options, and being accountable to the company for learning, progress, and results. Interestingly, the two companies that began our study with no specific group responsible for helping carry out the BI mandate developed these groups over the course of our four-year observation period. "If everybody owns it, no one owns it," is a common refrain that aptly describes the need for some organizational group to take this on.

The innovation function is tasked with making breakthrough innovation happen, so it has to come up with activities and programs and be identified in the organization as the group responsible for this mandate. All of this is infrastructure.

Activities and Programs. BI groups in companies engage in the following activities:

- Stimulate idea generation. They facilitate workshops, capture ideas from within the organization and scout outside the organization, and develop or elaborate them to the point where a project can be initiated.
- Engage senior leadership to influence and clarify strategic intent and thereby articulate domains considered ripe for innovation.
- Seed funded projects.
- Coach project teams.
- Scan the external environment (universities, other companies, government programs) for trends.
- Make investments in small firms or in venture capital funds to be at the forefront of particular technology developments.
- Interact with organizational support functions (human resources, finance, purchasing, legal) on behalf of the project teams to ensure those functions' support of the innovation world in addition to the operational excellence world.
- Help design the governance and oversight board and facilitate portfolio reviews.
- Engage in ongoing monitoring of the company's BI capability and recommend improvements.

So BI system staff take on facilitation, education, scouting, coaching, and some evaluation to initiate and seed-fund early projects. These activities and programs are all designed to stimulate and guide breakthrough innovation. They're necessary activities. Innovation doesn't happen on its own.

One caution: BI system staff can become overinvolved in doing the project work themselves when pressure starts to tighten to get one or two projects out the door. In addition, in the early phases of BI capability development, we observed that coaches also served as portfolio evaluators. This sets up an inherent conflict of interest, since innovation system staff who've coached specific projects may become biased in their ability to evaluate the project's promise compared with other projects in the portfolio. Some companies preferred this approach, indicating that they appreciated having a representative on the evaluation board with insight into the project beyond what the team could present. Others agreed that there were conflicts of interest in this setup and that the time spent with the teams became formalities rather than working sessions. These companies restructured their portfolio review boards over time to exclude project coaches.

Location and Reporting Structure. When most of the BI groups we studied were getting started, they reported to the CTO of the company. This is because R&D was the most concerned about the company's long-term innovation agenda. However, over time the organizations recognized the need for incubation and acceleration activities, and so, as we have described in previous chapters, the original organizations were expanded to incorporate those building blocks, and in many cases they were located elsewhere in the company.

With the full awareness of the DNA building blocks that we now have, firms do not have to begin at the beginning as our companies did. Rather, you may decide to build a discovery and perhaps incubation capability in R&D but house acceleration elsewhere. Only three firms in our sample were concerned with anything beyond discovery at the outset of the study, because the initial challenge was developing a flow of good ideas and getting projects started. But the others collected orphan projects in the company that had not gotten enough attention and for which everyone believed there was great potential. When you start with a pipeline of projects, you may find that incubation or acceleration is the building block that you need to set up first. You have the choice. It depends on where your biggest gap is. If you have projects with no support, start with incubation. If you have no projects at all, focus on building your discovery capability first. If you have lots of small businesses with big potential, it may be that acceleration is where you begin. Then backfill as needed to enrich the pipeline and infrastructure.

Start Working, Accumulate Experience, and Then Focus on Processes

Starting something new like a breakthrough innovation function can be exciting but also risky. Your peers know you're doing this, your management knows it's going on, and many in the broader company also know it's going on. The company has likely attempted this in the past and failed. So, yes, it feels risky.

Some of our BI leaders handled this by focusing on getting the system, processes, and infrastructure all set up before they started seeding any opportunities. Or they started by initiating calls for proposals and evaluating each one from scratch. Or they focused on educating others to change the culture to be more supportive of innovation. But they did not engage in innovation. They analyzed it, they talked about it, they reviewed some ideas—but they didn't start. This is the sign of risk aversion to the extreme.

At some point, you have to take the plunge. Get some projects going, or identify some major business opportunities and pursue them—and sooner rather than later. Learn your processes as you go. Figure out that you need additional staff as you grow. But get a pipeline going. Get started with some opportunities. It's through experience developing opportunities, nurturing them in an incubation mode, and experiencing the explosive growth of acceleration that true learning takes place. Analysis paralysis will get you nowhere in the world of innovation.

One of the companies in our study worked diligently and at length to clarify the processes for finding, evaluating, and initiating projects; managing projects; populating teams; funding projects; and monitoring them. Their drawings, flowcharts, and posters were masterpieces. But they had no projects in the pipeline. "We want to build the system first, then turn it on," they told us. They believed that once they issued the request for ideas, the flood of opportunities would require that they have a well-oiled machine to handle them. They spent months perfecting their policies and procedures. When they finally turned on the system, a trickle of ideas came in. Ultimately they had to go out and look for the projects that'd been around for awhile to populate their portfolio.

Overengineering this system isn't worth it. Get some projects that have interesting possibilities, and start. Allow the processes to evolve as you learn. You'll develop processes for all of the activities that you'll engage in:

- Finding and evolving ideas
- Screening and evaluating ideas
- Populating teams
- Working with and coaching teams
- Killing projects
- Interacting with markets
- Sourcing and managing funds
- Developing and managing internal interfaces
- Working with the finance, human resource, purchasing, legal, and merger and acquisition communities
- Working with business unit leadership
- Working with R&D leadership
- Governing the portfolio

- Developing education and culture change programs
- Continuous improvement of the BI capability

Trying to develop these processes before there is an adequate experience base is a waste of time. Once it's clear that the objective is breakthrough innovation and that the opportunities will be ambiguous rather than clear at the outset, the processes will evolve to meet those objectives. The risk of detailing processes before having the experiences is that you'll adopt the same processes that you are sure work in other parts of the organization but may not be the right ones for the innovation world.

Announce Your Existence

The participants in our study were surprisingly shy about this. In fact, a BI group in one company told us that they did not tell anyone in the company that their group had been formed. They wanted to underpromise and overdeliver. The idea was that they'd incubate a project or two. Then when they were gaining market traction and the fledgling businesses were becoming exciting, they'd draw attention to themselves based on those results. If they announced their existence, the logic went, the pressure to perform would be on. Heightened visibility increases vulnerability.

But breakthrough innovation doesn't happen in a vacuum. Any time the project leaders needed help from anyone in the mainstream organization, or resources, or attention, it was not forthcoming. Why not? No one understood the importance of their activity to the company or that the company had decided to legitimize this group. They lacked organizational credibility.

Let everyone in the company know who you are and what you are doing. Let them know you're developing a pipeline of new business opportunities and welcome their ideas. Just as other functions in the company call for assistance or need internal partners to accomplish an objective, so too will the innovation group. The announcement should clarify the group's role and set expectations accordingly (longer-term, bigger hits). But beware: this will take time on the part of the orchestrator and the innovation function leader. It's a big investment of time and energy. Communicate, communicate, communicate.

The announcement should help establish the language of innovation in the company (what vocabulary will your company use for breakthroughs?). There are new programs, new advisory boards, and new roles, and these should be clarified. All of this lends legitimacy to the innovation agenda for the company.

When Johnson & Johnson Consumer Products CTO Neal Matheson called all of his people to the atrium of the research building in early 2000 and announced that breakthrough innovation was of keen strategic importance to the company, he described to everyone the programs he had underway with his vice presidents to make it happen. His endorsement legitimized all of the activities they were undertaking. When IBM announced its EBO program at two large management briefings in Armonk, New York, and Tokyo, Japan, in early 2000, and named its EBO vice chairman that summer, everyone in the company understood that group's role and could decide how they might interface with it. Mike Giersch, vice president of strategy, told us, "We did not wait to announce this. This was a very important thing. We needed people on board to support the effort." Similarly, GE's announcement under Jeff Immelt of its advanced technology programs was widely communicated internally and externally.

It's true that this puts some pressure on the group to perform. But they should be pressured to perform, as everyone else in the organization is. The important point is to ensure that the performance expectations that you set are realistic given that you are not in the business of traditional new product development but rather in the more uncertain world of new business creation.

Explore Resource Models

We observed many funding models for BI activities in the course of our study, and the surprise was how poorly clarified they were in many companies. Many of the companies' leaders called for break-through innovations, but expected those in charge of the innovation group to use their current budgets to fund it along with their normal funding responsibilities! Other companies did provide a budget to fund the innovation group and get projects started (seed funding), but they had no plan for how to handle growing opportunities requiring big investment dollars. Still others planned a cost-sharing program with business units as the fledgling innovation businesses matured but did not have an answer for how to fund unaligned opportunities.

At the beginning, this may not be a problem since the BI group usually focuses on finding discovery-based projects and seeding them with small bits of funding in order to develop a pipeline. But when the projects begin to grow and require more resources, or the group's seed funding budget is fully used, the issue of funding models becomes heightened. At what point should business units begin supporting or sharing in the cost of these businesses? Will they be forced to pay or allowed to pay only after all the risk is removed, as in the acquisition funding example we reported in Chapter Five? What about unaligned opportunities, which will not benefit any business unit, or multialigned opportunities, from which several business units will benefit but none is willing to fully fund? Furthermore, how much money should be granted in seed funding to a project team? One of our companies that wanted to stimulate innovation funded nearly every initial opportunity that was proposed—but gave each team too little money to allow much progress.

Human resources are also an issue. When people are needed on project teams, how will you access them? Will the BI group pay for their time as a part-time project team member, as the Shell Chemicals GameChanger group did, or will the innovation budget employ a large number of individuals who do all the discovery and incubation work on all the projects as Kodak's SCC did? Will the group expect that the expertise in new business creation will be added from a relevant business unit, as is the case in GE, with its business program managers? Or will there be expertise in new business creation resident in the innovation group to coach the project teams but not join them, as is the case with 3M's business development group and Corning's exploratory marketing team? There are many different models, and they all have advantages and disadvantages. These issues must be surfaced and decisions made as the innovation effort begins. Then monitor to see if it's working for you. If it's not, adapt. And adapt quickly.

Keys to Getting Started

The point here is not so much to provide answers, but rather to raise relevant questions to address. Answers to these questions vary company by company and as the BI initiative begins to be developed. Here is a summary of a number of key points for getting started:

- Address each of the elements of the management system at the outset. Don't ignore any.

- Get the structure, strategic framework, and talent for new business creation in place first. Then build new processes as the company learns. Many companies build the processes first because this is what they are more comfortable with, but they have insufficient knowledge of what it takes to build a mature BI capability, so they do not build the right processes.

- Realize that you may have to spend time on education and culture change. Make sure this is built into your metrics at the beginning.

- Develop a language for breakthrough innovation that is legitimized and different from typical organization-speak. Words like *experiment*, *explore*, and *twenty-five-year plans* are okay.

- Develop a mechanism whereby a firm's strategic intent and BI system are mutually influenced.

- Develop a systematic method for capturing, sharing, elaborating, and screening ideas, for thinking about the future and considering opportunities.

- Establish a pipeline of projects by collecting some existing ones that may have potential but have fallen through the cracks due to poor fit or inability to apply appropriate processes. In this way, you'll have some businesses graduating within the first couple of years and can avoid pressures to perform.

- Clarify the role of the breakthrough innovation initiative in terms of the company's future.

- Bring in people from outside the company who have had business creation experience to work as part of the team.

- Ensure the leadership of the effort has business creation experience as well as rich, powerful networks inside the organization.

- Develop processes that are appropriate to exploration, experimentation, and hypothesis testing to jump-start the incubation competency.

- Figure out where the money and people resources will come from. You may need to experiment with different approaches, but don't allow senior management to suggest that a BI capability is needed but then not allocate any resources to it.

- Consider the BI system as having three building blocks (discovery, incubation, and acceleration) that must be tightly linked but require different competencies. Start to build the block that you need most immediately. If you have ideas, focus on incubation. If you need ideas, focus on discovery.

The Maturing Breakthrough Innovation Capability

After some period of time (usually a couple of years), the innovation group is known in the organization and has developed a pipeline of projects but likely has not yet commercialized anything, especially if it is starting from scratch. Even if it started with some orphan projects, some of them will have died on the vine. Patience may be wearing thin. The group may still be struggling to find the right staff and project leaders and may perhaps be fighting for a bigger budget as they realize the need to add incubation or acceleration capabilities, or both. That's what we saw in the majority of our companies. Ideas are flowing, but they're too small or way too big, nothing's turning into businesses, or they're stepping on others' toes in the company. These are the challenges!

But you should know that it's not all challenges. There are great successes along the way. BI teams are beginning to understand and feel more confident about their investment decisions and ability to populate project teams appropriately, and they're seeing several of their portfolio opportunities making interesting progress. Innovation leaders are generally pleased with the core staff they've assembled by this time and are encouraged by the progress they've made in generating interest and enthusiasm in innovation throughout the company. Still, their enthusiasm is dampened by their anxieties.

Figure 8.1 highlights some of the major challenges we saw. Even if there's commitment from the top, it's difficult to face the long time horizon that is required to create a new function from scratch. We begin to see cracks in the system after a couple of years.

We describe these challenges a bit and then give you a perspective on how to handle them. Most of this perspective comes from the fact that not all companies in our sample reacted similarly to the challenges, so we saw some reactions that worked better than others. Now that we know what the whole system looks like, many of these are avoidable at the outset.

Idea Generation and Scope

Companies that started building their BI capability by issuing a call for new ideas soon recognized that individual projects won't get them to their objectives of breakthrough change. "We're a rounding error to the overall company's top and bottom line," we were told. They eventually all moved toward a platform approach—either technical platforms or technical-business platforms. Those

Figure 8.1 Characteristics of a Maturing Breakthrough Innovation Capability

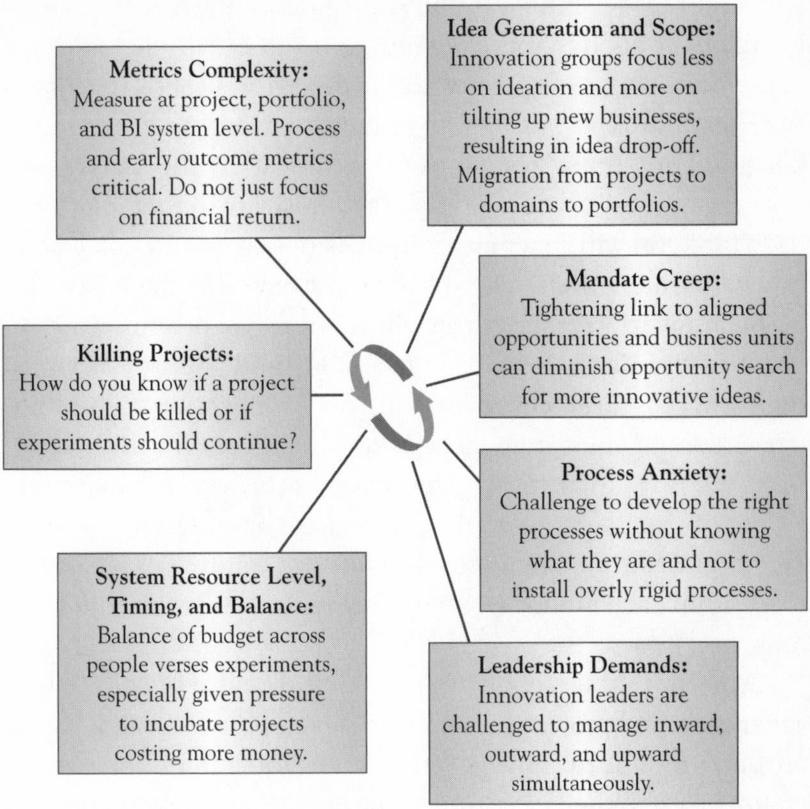

Metrics Complexity:
Measure at project, portfolio, and BI system level. Process and early outcome metrics critical. Do not just focus on financial return.

Idea Generation and Scope:
Innovation groups focus less on ideation and more on tilting up new businesses, resulting in idea drop-off. Migration from projects to domains to portfolios.

Killing Projects:
How do you know if a project should be killed or if experiments should continue?

Mandate Creep:
Tightening link to aligned opportunities and business units can diminish opportunity search for more innovative ideas.

Process Anxiety:
Challenge to develop the right processes without knowing what they are and not to install overly rigid processes.

System Resource Level, Timing, and Balance:
Balance of budget across people verses experiments, especially given pressure to incubate projects costing more money.

Leadership Demands:
Innovation leaders are challenged to manage inward, outward, and upward simultaneously.

that began this way at the outset have commercialized bigger impact businesses more quickly. A platform is a technology domain or a business domain (or both) that enables multiple products and, indeed, product lines or product families. Analog Devices' accelerometer technology platform, developed in the late 1990s, enabled it to develop business in the automotive, gaming, medical equipment, and many other industries. IBM's pervasive computing EBO has launched many new service opportunities for that firm. GE's move into biotechnology is launching many new businesses in the field of information-based medicine.

This "big hit" platform orientation also encourages, and actually forces, the conversation between the innovation group and senior leadership regarding strategic intent, thereby increasing the likelihood that the organization will be willing to receive and embrace the outputs of the innovation group. It reduces the organizational uncertainty dimension that we described in Chapter Four. It also encourages thinking along the lines of longer time frames and of building portfolios within each platform. Each platform will generate a variety of opportunities, some more near term than others, which will help strengthen the legitimacy of the innovation group within the mainstream organization and begin to stimulate the market to understand and interact with the technology, thereby identifying new applications. The challenge that our companies faced is how to prevent the entire project team from focusing on the immediately commercializable aspect of the platform while neglecting the longer-term prize. Teams must orient themselves to harvest small wins along the way within each platform while continuing to invest in the long-term prize.

And that's the second big issue that surfaced regarding idea generation and project initiation in developing BI functions. As pressures mount to show progress, some companies' innovation groups diverted their attention from generating new opportunities and focused heavily on working on those they had in the pipeline. Eventually most realized they had to reenergize their discovery focus. We note that consistency within the discovery team is important. When the same people review ideas, they develop a collective memory of what they've already seen. They spot trends this way and can identify combinations of ideas that can suggest interesting business platforms. So consistency of the idea review group is something we suggest. This is counterintuitive, because there's a need to ensure freshness in the idea generation activity. The ideas, we note, can and should come from everywhere. It's their elaboration, review, screening, and articulation that should be handled by a consistent team.

Mandate Creep

Companies whose BI mandate was to find opportunities in the white spaces tend to have projects that are more aligned with the business units. And with that, in several cases, the mandate slowly shifted from breakthrough projects to those that are more incremental. The reason is that they cannot get organizational acceptance for projects that have no clearly identifiable home, particularly if senior-level governance boards are not in place to trumpet the importance of white space growth and encourage it. In addition, BI innovation staff may be applying traditional skill sets and conventional processes to the opportunity, and they gravitate toward the comfortable. Or they're underresourced and so tend to rely on current customer connections and current technological solutions, which draws them into the "breakthrough ideas, incrementally executed" trap we described earlier.

Companies in our study whose BI mandate was to find aligned opportunities noticed that many opportunities they come across are unaligned, in the white spaces between their current lines of business. This occurred for two reasons. First, when they were searching for opportunities, undoubtedly these inconveniently unaligned but very exciting opportunities appeared. Second, BI groups found they needed to expand their scope as they realized that breakthrough ideas in businesses that the firm has been operating in for years (aligned opportunities) are few and far between. The action is at the interfaces between business domains. That space opens up numerous new dimensions for innovation, and that's where most innovations tend to occur.

DuPont's APEX system team noted a creep in its mandate toward more aligned and nearer-term opportunities. Things were tough in the company, and the pressure was on R&D's APEX system to generate big hits quickly. No one was pleased with the success rate in the portfolio and believed that the churn rate of projects was too high. Once the team realized they were migrating to closer, aligned opportunities in order to increase their success, they became concerned. They decided to turn up the pressure on themselves and increased their discovery research budget. In addition, they decided to expand their inbound marketing staff's roles beyond opportunity generation to include project incubation as well. They checked their mandate creep and decided it was not what they intended.

Process Anxiety

The BI leadership team can begin questioning whether the processes they've adopted are the appropriate ones as they see project teams and innovation coaches struggle. Companies aren't great at operating in the realm of uncertainty, yet uncertainty is part of the challenge and fun of innovation. Getting used to it may be difficult. Trying to contain it by developing rigid processes is a natural tendency in organizations. BI leaders asked themselves:

- How do we measure project progress?
- What are the appropriate ways to manage projects whose plan is constantly disrupted by new learning?
- How much time should be spent in the field?
- What should we expect from our exploratory marketing endeavors?
- How do we set up partnerships appropriately for BI teams?
- How do we get better at evaluating ideas so we can fund fewer bigger projects?

One incubation leader told us, "We're laying the tracks as we're driving the train!"

We have described a number of principles in this book that can help BI groups develop and adopt appropriate processes: the learning plan for project management; criteria and methods for evaluating portfolios; descriptions of roles and responsibilities for innovation group members; and descriptions of activities needed in discovery, incubation, and acceleration to which our participating companies did not have access. Indeed, they have been charting the course on new processes. We note two important principles for companies embarking on building a BI capability based on our study results:

- *Don't let processes own you.* Remember the group that told us, "We need guidelines, not process"? They were absolutely right. Every new business initiative will be different. Each will require the company to venture into new technical areas, chart new markets, and perhaps develop new business models. To assume that you can order this into a series of steps that can be replicated each time would be innovation suicide. If it's replicable, how is it innovation? Each year, Mike Giersch and IBM's EBO leadership team coordinate a series of exercises to generate new ideas. He and his staff were considering new ways to do this and were analyzing which method that they'd experimented with was the most effective. They came to the conclusion that changing the method each year was one of the reasons they were coming up with unique ideas. They were reaching into different pockets of the organization, using different techniques, and considering different methods. Those differences, in and of themselves, are valuable for idea generation.

- *Remember the innovation group's role in the organization. Let that drive process development.* The role of the BI group is to explore and experiment, and then exploit. Your processes will not be the same as those of the operational part of the organization. Our companies struggled with how much to conform to the organizational processes and vocabulary and how much to distinguish themselves from those processes. Some told us they used loose stage-gate processes. Everyone was supposed to understand what that means. Others, however, developed their own and worked very hard to ensure the company understood the distinctions. The latter approach requires heavy investment in education, communication, and perseverance to develop processes that are appropriate and work. If innovation is going to be a true function in the company, it should have a vocabulary and set of practices by which it is distinguished and support its mandate in the company.

Leadership Demands

Within a short period of time, the complexity of the BI group leaders' roles becomes clear. They must simultaneously manage upward (to senior corporate leadership), outward (to business unit leadership), and inward (their staff). There is a heavy communications responsibility to all stakeholders in innovation activities.

Managing Upward. Managing upward involves reporting on progress, influencing corporate strategic intent, and continuing to educate the leadership about the impacts and importance of innovation and how it is properly managed. This is exacerbated by the fact that senior leadership turns over frequently in many companies. Seven of our twelve primary participating companies experienced CEO or CTO changeover (and five of those experienced both) in the four years of our observation period. Each time, the BI group was vulnerable. It was up to the group leader to convince the new senior managers of the need for an innovation group, its reasons for existing, potential benefits to the company, and why investment in building long-term businesses based on breakthrough innovation was critical to the company's long-term success. Some innovation groups handled this vulnerability by communicating about the company's investment in building a BI capability to external people, such as the media and market analysts. They viewed having the innovation group legitimized outside the organization as an important way to maintain their foothold internally during times of leadership change. This is another reason that innovation must become a function rather than a program in companies. Surviving leadership change is a difficult task and consumes more time than BI group leaders have to spare.

Managing Outward. In terms of managing outward, a host of issues arose for the companies we studied. First, as the company becomes more broadly aware of the BI group's existence, the business units began requesting help from the group for new product development problems they could not solve themselves in some of our companies. The innovation team must ask itself if this sort of consulting role is part of its mandate.

Second, business units may have set up an innovation group of their own for aligned breakthrough opportunities. As the complexity of the innovation system evolves or elements of it experience change in leadership, lack of interfaces may occur for a period of time that create tension. This can happen within the innovation function itself (for example, miscommunication or overlapping responsibilities of discovery, incubation, and acceleration) or between the innovation function and other parts of the company. In one of our companies, the innovation group's mandate was to develop opportunities in the white spaces. The business units were supposed to be responsible for their own aligned opportunities. The innovation group migrated toward more aligned opportunities over time because the company did not have an acceleration capability in place to house growing unaligned businesses, causing redundancy in the organizations. Ultimately the innovation group reverted to its original objective of unaligned opportunities and developed a mechanism for fast-tracking the aligned opportunities to the business unit early and quickly.

In a number of the companies we studied, projects were incubated in the business units, so the issue of a transition to business units at that early point in their development and deciding how to maintain some sort of oversight arose. In many cases, this involved politicking for business units to be measured on long-term-growth initiatives as well as short-term profitability. Otherwise it's difficult for the innovation group to get the attention and resources needed for those projects to be properly incubated. Funding models and cost sharing with business units also had to be negotiated and renegotiated.

Finally, heavy investment of BI leadership time was required to develop and maintain relationships with corporate staff functions whose help was needed. Access to legal, financial, marketing, purchasing, manufacturing, and human resource support was constantly cultivated. It wasn't so much that these groups refused to grant support, but rather that they didn't know how to help. Legal agreements needed to be written more loosely, hiring practices had to reach more broadly and into unusual domains, and purchasing of one-off pieces of equipment had to occur with almost no notice at times. In essence, many of the rules and procedures that help streamline and protect an ongoing operation became hurdles or barriers for the innovation function. Working to secure the kind of help they needed, rather than to ignore the corporate functions and handle these issues on their own, consumed lots of BI leaders' energy in many of our companies.

Managing Inward. Managing inward poses challenges of talent management, role clarity, skill development, and attitude management for those precocious individuals who can be so valuable to an innovation organization. While some of these issues have already been raised in an earlier chapter, others require elaboration.

One of the challenges facing BI leaders is the recognition that those who stepped up to the plate when the innovation group was forming may not have the right skills and background to participate or manage the innovation activities for the long term. In addition, the team as a whole may not have the full complement of skills needed to nurture projects, manage the portfolio, handle education, feed the pipeline with ideas, run interference throughout the organization, and make connections to necessary external constituents on behalf of the project teams. The initial focus on culture change and education may stall this recognition for a period of time, but it becomes abundantly clear as the pipeline fills and projects are not progressing.

We saw many technically trained people trying to coach project teams, and the outcomes were predictable: lots of technical advice but no pushing, prodding, or help with getting into the market or validating a business proposition. These issues can stall a project far too long and promote inefficient use of time. The BI group needs the complement of analytical people, technical advisers, exploratory marketing and business development people, and strategic thinkers for taking ideas from discovery, through incubation and acceleration, to commercialization. These encompass hard skills such as technology development, market analysis, and developing strategic clarity for each business as well as soft skills to help negotiate relationships, build networks, and find and replace BI team members as necessary.

BI leaders complained to us several years ago that they could not offer their staff an innovation career path for all the work that we've described associated with DNA activities. This resulted not only in demotivation and loss of people from the group who were concerned with upward mobility; it also caused stagnation and lack of independent thinking over time for those who remained.

There was also no formal mechanism for hiring people into these groups from the outside, so innovation leaders obtained the necessary skills in nontraditional ways: contracting with external inventors, hiring student interns from M.B.A. programs with a focus on innovation and entrepreneurship, hiring "opportunity brokers" to assess external interest in the early business opportunities, and leveraging the founders of start-up firms that the company had acquired. In one case, the BI leader issued a widespread call for help to all members of the organization taking M.B.A. courses in evening or executive programs. He noted that he could supply them with fodder for many of their marketing or entrepreneurship course projects. These are creative acts, and we applaud the innovation leaders for being so entrepreneurial themselves, but they're acts of desperation due to the lack of a systematic approach to building this expertise in an organization. Companies can do better, and they are. Several of our companies have launched formal efforts to develop human resource strategies for recruiting, developing, and retaining those in the innovation function.

The Complexity of the Leader's Job.   All of the issues that BI leaders faced as their functions became more mature are certainly manageable, but they demand energy, time, and attention. In the long run, we may find that these issues might be the core of an innovation group leader's job description.

System Resource Level, Timing, and Balance

BI group leaders expressed challenges over time in determining how to balance their resources. Several indicated they'd spent too much on people and needed more resources for experimentation, equipment, travel, and prototype development as more of their projects needed incubation activities. In addition, the need to work around traditional budget cycles was a challenge. The BI portfolio may contract and expand as opportunities arise and are paced through discovery, incubation, and acceleration, yet the traditional budget model is incremental change year over year. Balancing the portfolio across the building blocks may help even the expenditures each year, and so align the timing of resource needs of the innovation system with that of the operational system.

Killing Projects

If everything is working as it should, project teams will let you know when a project should be killed. If the fear of failure has been overcome, as it should in the situation of breakthrough innovation, then teams will tell governance boards when they've hit too many dead ends in a project and cannot make progress. When a backlog of projects exists (as we described in Chapter Three with the example of DuPont's practice of having a bench inventory of projects waiting for teams to tackle), the tension of killing projects and the difficulties associated with reallocating people to new projects are lightened, making it easier for teams to admit the need to move on to bluer oceans.[3] This occurred in several of our companies, but for many, killing projects was a real challenge after they'd been building the portfolio for some period of time. Remember, fast failure is your friend.

Metrics Complexity

BI leaders may find that the metrics originally considered appropriate actually are not. One of our companies was measuring its portfolio on the basis of top-line revenue but realized over time that this drove the innovation group to keep projects too long so they could take credit for those revenue streams longer—and perhaps avoid growing new ones. If measured on the number of transitioned projects (the most commonly used measure we observed), there's a tendency to move projects into the next portfolio (incubation, acceleration, or operations) too soon. All of these lessons were learned through experience. Ultimately most companies realize that the innovation system should be monitored at three levels: projects' progress, the portfolio (as described in Chapter Six), and the innovation system, as summarized in Table 8.2.

Systematic reflection on the last set of issues is key for building any new competency or effecting any change in an organization, and innovation is no different. BI leaders we studied did appear to invest in reflection, discussion, and purposeful change of processes and infrastructure over time. They'd read widely about innovation management (several of them even teach it), hire consultants, reflect with their group, and make changes accordingly.

Conclusion

What if you knew all of this ahead of time? Well, now you do. So what are the implications?

- Don't start from the beginning. Start from the middle. Collect orphan projects, and incubate them first. Build discovery as needed to enrich your pipeline.
- When you think about discovery, think about strategy, big platforms, and businesses rather than products. Each platform should be generating myriad opportunities. Each one is a portfolio in and of itself.

Table 8.2 Three Levels of Metrics for Breakthrough Innovation

Level of Measurement	Appropriate Metrics
Project level	Will vary if in discovery, incubation, or acceleration
	Richness or promise
	Clarity of strategy
	Closeness of fit with where we want to be
	Whether there are executable models
	Whether the market shows enthusiasm
Portfolio level	Portfolio size in terms of number of projects
	Portfolio scope in terms of game-changing potential
	Portfolio diversity across dimensions of importance (domains, pacing)
	Portfolio activity: churn and transition rate
System level	Individual elements: Are the elements of the innovation management system working as they should?
	Alignment within: Are the elements of the management system aligned with one another?
	Linkages within: Are the interfaces within the innovation system working as they should?
	Linkages to mainstream: Are the interfaces between the innovation system and the rest of the organization effective?
	Is the innovation system having the impact on the company that we desire?

- As you build your team, think about the kinds of people you need. Motivation is worth a lot, but experience and acumen are critical.

- Communicate, communicate, communicate. Keep rationalizing the innovation group's role in the organization in relation to other parts of the organization. Continuously revisit the mandate and recheck commitment.

9

THE INNOVATION FUNCTION

Since we began our research in 1995, and particularly since we began the second phase in 2001, we've been struck by the attention and progress that firms are making in building capabilities for breakthrough innovation. There are many signs that BI capability development is taking root in established companies. Firms are recognizing the need to distinguish between R&D (the engine of invention) and innovation (leveraging invention to create new businesses). They're also beginning (but just beginning) to recognize that breakthrough innovation must be managed differently from new product development or process improvement. Groups are springing up in companies that have intriguing labels: Future Business-Creative Center (Bayer Materials Science), Global Innovation (MacNeil), Innovation Strategy group (Millipore). Corporate New Ventures (P&G), Strategic Growth (Corning), and Corporate New Business Development (Sealed Air), to name a few. But it's definitely still new. Companies are experimenting with new processes and new organizational structures; they're asking about appropriate metrics and beginning to wrestle with portfolio-level issues. But they're not backing away. It's a period of foment right now. Only two of the twelve companies we studied in phase II reduced their commitment to innovation over the four years we studied them. And of the remaining ten, several have indicated to us that the business units are now competing for the BI group's attention and resources. They view the group as important to their long-term success.

Looking Toward an Innovation System

Many of our participants tell us they believe they're on the right track but wish they had better direction. The experimentation with management practices associated with BI means to us that companies are viewing this as an area to learn about and improve on. Senior leadership tells us that they may not know how to do it, but they know they need it. Breakthrough innovation is the next management frontier for companies; it's not a fad. More and more companies are interested in building a sustainable breakthrough innovation capability, which they view as essential to their near-term health and long-term survival.

We held multiple workshops with our participating companies throughout the course of our study. At the first one, the theme we heard loud and clear was this: the inconsistency of commitment from senior leadership is the showstopper. So how does a BI capability move from being a program, or a pot of money that gets resourced occasionally, to a sustained aspect of the company's daily activities? The answer is that innovation has to, and is, becoming a professional business function. Just as marketing emerged in the 1950s from being a set of activities that was funded or not at the discretion of the leader to a full-fledged ongoing function, complete with leadership, metrics, its own vocabulary, career paths, and staying power in organizations today, so too will innovation.

A full-fledged innovation function is a corporate-level function with reaches into a company's divisions if the firm is large enough. Just as marketing groups consist of market research, marketing communications, sales organizations, and perhaps others, so too does the innovation group encompass multiple responsibilities. A BI function is responsible for building and nurturing a portfolio of innovation opportunities; overseeing the health of the discovery, incubation, and acceleration activities; and orchestrating the relationship with the mainstream organization, with the understanding that that relationship will constantly change as the company's capacity for innovation ebbs and flows.

We observed this model, in various stages of development, in seven of the twelve companies in our primary study group and validated it in the remaining nine companies of our validation sample (see Preface for description). It's this model that we've fleshed out in detail in these chapters and summarize in Table 9.1. Once you've become familiar with the principles of an innovation function, you can adapt the practices to fit your own organizational capacity. But one thing's for sure: the management system for innovation must exist in order to support any desire your company has to make breakthrough innovation happen on a consistent basis.

We have discussed at some length in various locations throughout the book that breakthrough innovation is shrouded in deep levels of uncertainties in many dimensions. Therefore the management of breakthrough innovation requires multidexterous capabilities of key people in a variety of dimensions. These people must exercise their internal and external networks for learning, and employ astute experimental savvy to rapidly reduce uncertainties.

But uncertainties do not necessarily correlate with risks. It is common to believe, or at least assume, that breakthrough innovation efforts carry with them extreme levels of risk. But this is far from being so. The management system for innovation described in these pages is a system for reducing uncertainty of breakthrough innovation and coping with project risks. Rather than betting the company on a once-per-decade possibility, companies can imbue their organizations with an innovation capability.

Risk is a two-edged sword. Although there is risk, we have ascertained that the long-term risk to the institution from not having a strong innovation capability exceeds, and possibly vastly so, that of engaging in innovation investments.

A Management System for Breakthrough Innovation

In Chapter Eight we described steps along the path to developing a mature or sustainable breakthrough innovation capability. We based our description on the experiences of companies in our study.

Table 9.1 Overview of the DNA Breakthrough Innovation System

	Discovery	Incubation	Acceleration	Operations	BI Systems
Primary activity	Explore	Experiment	Escalate	Exploit	Integration and coherency
Mandate and responsibilities	Create a pipeline of possible opportunities and clarify them in accordance with company strategic intent.	Nurture projects through vetting technical, market, resource, and organizational issues to understand how opportunities may play out as businesses.	Build to maturity high-impact businesses to more predictability and acceptability to the mainstream organization.	Leverage an innovation business platform into numerous long-running successful offerings.	Manage composition, progress, and flow of portfolio of potential high-impact businesses in symbiotic fashion with the mainstream organization.
Organizational structure and process	*Structure:* Centralized and tightly linked to R&D. *Processes:* Constant development of internal technical capabilities, external and internal scanning, and open sourcing of ideas. Build proficiency in opportunity elaboration and articulation through	*Structure:* Dedicated unit in corporate or business unit and tightly linked to R&D. *Processes:* Bench inventory of projects to help kill less valuable opportunities. Learning plan for managing high-uncertainty projects. Experiments on	*Structure:* If focus on aligned, then in business unit. If unaligned or multialigned, then special unit in corporate. *Processes:* Manage for high growth. Focus, respond to market inquiries, invest in demonstrating a path to profitability.	*Structure:* Business unit. *Processes:* Developing next-generation products and new applications using stage-gate processes. Focus on quality and efficiency using Six Sigma.	*Structure:* Dependent on innovation capacity: centralized structure incorporating all building blocks, or disaggregated but linked building blocks. *Process:* Build DNA awareness, and monitor and diagnose innovation management system health.

	socialization and ties to experts. Identify, combine, filter, and select opportunities for further development. Maintain inventory of projects to help kill less valuable opportunities. Develop opportunity flow.	technical, market, and organizational uncertainties. Use coaching to help clarify each business's emergent strategy.			Manage the churn rate in accordance with objectives.
Resources and skills	*Resources:* Corporate-level resources. *Skills:* Motivated individuals. Creative, insightful, inductive reasoners with a penchant for strategic thinking.	*Resources:* Dedicated budget but access to external pools of resources for projects when possible. *Skills:* New business creation expertise, staff with coaching and nurturing capabilities. Team leaders selected for entrepreneurial acumen and rich networks.	*Resources:* Top-level investment, primarily from within the corporation unless the business is a joint venture. *Skills:* Acumen in nurturing high-growth businesses. Ability to interface with mainstream organization.	*Resources:* Gained through budgeting processes. *Skills:* Personnel associated with operational excellence, capital funding models, and other traditional businesses.	*Resources:* Human and other corporate resources. Commitment of people and access to funding and office and laboratory space. Senior leadership time and attention. *Skills:* Systems thinkers, negotiation, influence, motivational and networking capabilities.

(Continued)

Table 9.1 (Continued)

	Discovery	Incubation	Acceleration	Operations	BI Systems
Leadership, culture, and governance	*Leadership:* Typically owned by CTO and managed by middle-level manager. *Culture:* Fluid, imaginative, and collaborative. *Governance:* Experienced evaluation board.	*Leadership:* Need an incubation leader; could be filled by CSO, CTO, or vice president of new business development. *Culture:* Inquisitive and learning oriented. *Governance for projects:* Project advisory boards with expertise appropriate for the opportunity. *Governance for portfolio:* Portfolio governance board.	*Leadership and governance:* Senior leadership team with powerful networks, respect, and political clout. *Culture:* Hard driving, urgent.	*Leadership:* General management business or group leaders. Middle management may be functional leaders. *Culture:* Operational excellence. *Governance:* Hierarchical business unit structure.	*Leadership:* Corporate officer oversees system and is ultimately responsible for innovation, aided by orchestrator. *Culture:* Strategic orientation and passion for innovation process. *Governance:* Senior leadership team.

Metrics and rewards	Quantity of ideas and richness and robustness of concepts.	Milestones at the project level, churn rate, and internal rate of return at the portfolio level. Magnitude of opportunities (platforms over projects). Spillover of learning to other parts of the company.	Growth in sales and inquiries of portfolio businesses, identification of migration path, and uplift and spillover to businesses and BI activities.	Revenues, margins, stockholder returns.	Successful transitions of BI opportunities across DNA and into mainstream organization. Impact of BIs in the market and return on investment of BI portfolio investments over time.
Critical problems	Too many little ideas. Not enough flow into the BI system	Big ideas incrementally executed. Mandate creep. Lack of patience and appreciation of a learning orientation.	Lack of integration with mainstream organization. Transitions to business units too early. Not willing to put in time, effort, resources, and money to build at an appropriate speed.	Inability to accept BI into business units.	Developing awareness of DNA system and its dynamics. Lack of portfolio management.

Figure 9.1 Elements of a Sustainable Breakthrough Innovation Function

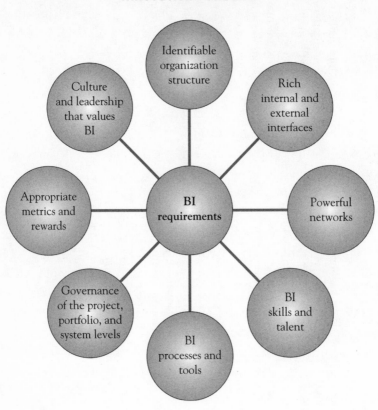

You can learn from those and leapfrog some of the challenges they faced by being aware of the opportunities and challenges required for a sustainable innovation function. The elements are summarized in Figure 9.1.

Identifiable Organization Structure

A sustainable innovation function needs an identified organizational group inside the company with an innovation-supportive culture. The group should be tightly linked to R&D and to the corporate strategy group. Some companies tell us that "innovation is part of the culture here . . . It's in our genes; everyone does it."

And yet we see that no one is tasked with the responsibility for making sure it happens. Although innovation may occur under those conditions, it's usually not breakthrough. Major innovations cannot be expected in an organic environment, where flexibility, consensus building, and fluidity are the primary managerial mechanisms for accomplishing objectives. Rather, repeated breakthrough innovation requires structure and clear reporting relationships to ensure the opportunity for both discipline and creativity. A clear set of roles and responsibilities is needed to sustain attention and resources to support the transformational experiences that constitute breakthrough innovation. An identified group can ensure this constant attention. In addition, an identified organization, staffed with people who are responsible for building and shepherding the BI capability and measured on its results, is more likely to reflect on its progress and improve its practices. Large, established companies offer room to learn and experiment that start-ups cannot afford. Without an identified organizational group, the practices that any single BI team finds useful will not be handed over to other teams. And the organizational practices that are used in the operating units will certainly overtake any experimentally based practices that a single innovation team may use.

Rich Internal and External Interfaces

A successful innovation function requires rich interfaces to the internal parts of the organization and external opportunity sources, funding sources, technical sources, and market access sources.

Internally the innovation group should be tightly linked to the company's strategic intent but loosely linked to mainstream operational processes and resource allocation mechanisms. The logic of the mainstream organization is based on stamping out variation, while the logic of the breakthrough innovation system is based on creating and increasing variation for learning and experimentation in order to increase the options for new businesses. The two systems should work in parallel.

The link to strategic intent is what drives the big investment decisions that breakthroughs require at some point in their development. If the purpose of BI is organizational growth and renewal, the link to the firm's vision for its future must be intimate and reciprocal. In other words, the reason for building a BI capability is to help the firm renew itself, with the directions for renewal executed through the BI system. Many new ventures divisions of years past began developing opportunities that were of little strategic importance to the company and were eventually disbanded. Case studies show that physical separation at the project level may work for a time, but complete separation at the system level may not be wise given that the purpose of an innovation function is to leverage and stretch current competencies while simultaneously building new ones. Thus, the interface to the mainstream organization at the level of developing and executing on a strategy for the company's future health becomes critically important.

By "loosely linked to mainstream operational processes," we mean that each system (innovation and operations) should be aware of one another's processes and resource allocation systems, so they can be connected for any given project when the time comes, but they should not be constrained to follow those. Some links may boost the likelihood that new business opportunities transferred from the BI system will be accepted. Eventually a fledgling business will need to be embedded in an operational unit and work within a business unit's systems. A project must at some point be added to the business unit's portfolio planning and resource expenditure plans, for example. Similarly, resources will be needed during a project's development that may already reside in an operating unit if the opportunity is tightly aligned with that unit. At Air Products and Shell Chemicals, for example, the BI system directors maintain a talent database: an inventory of people throughout the company they may need to draw on from time to time. Firms use idea-sharing fairs, technology conferences, and innovation fairs to loosely network members of the innovation system and mainstream operations so that mainstream employees are aware of innovation activities.

We find that adopting a unique vocabulary for the innovation system helps strengthen the strategic linkages while simultaneously distinguishing the processes for innovation from the processes for ongoing operations. When IBM declares a program an emerging business opportunity, for example, everyone knows what that means and what processes will be used to nurture it, fund it, and decide whether it's successful.

Powerful Networks

The innovation function must develop powerful networks inside the company. By "powerful," we mean people of influence over strategy and resources who can convince others of the importance of innovation in general and of specific directions that the innovation portfolio is taking in particular. The importance of external networks is well known, but in larger companies, the talent, knowledge, and resource bases can be vast, so internal networks turn out to be of key importance. Indeed, one of our recommendations concerns selecting managers in the BI system who are embedded in rich internal company networks and skilled at developing new networks as the need arises.

Breakthrough Innovation Skills and Talent

The innovation function must work to attract and develop a core staff with appropriate skills and talent. Those in discovery, incubation, and acceleration require different strengths, but these are all people interested in being pioneers rather than implementing someone else's plan. They are creative problem solvers, they are learning and doing oriented, and they love charting their own course. It is atypical to find people with these characteristics in established firms. Most have been driven out. Developing a reward and promotion system for these people is one of the biggest challenges facing firms.

Breakthrough Innovation Processes and Tools

Innovation-specific processes and tools must be adopted. Elevator pitches, learning plans, business proposals, discovery-driven plans, and real options-based evaluation mechanisms are key tools that help guide milestone-based learning, experimentation, and decision making under conditions of uncertainty rather than provide the step-by-step processes that assume a linear, predefined path and incite budding innovators to rebel.

Governance at the Project, Portfolio, and Innovation System Levels

A successful innovation function needs appropriate governance and decision-making mechanisms at the project and the portfolio levels. The innovation function itself is governed and overseen at a senior level. It may be the chief strategy officer or even a CTO that has business creation experience. Many companies are naming chief innovation officers (CNO), which we see as a positive sign. Expertise-specific boards composed of members internal and external to the firm are effective guiding bodies for each fledgling business. Portfolio-level board members must be company leaders who have the organization's long-run interest as their own and are willing and able to live with high churn rates and failure rates that may be similar to what a venture capitalist experiences. The portfolio governance board needs mechanisms to handle aligned, unaligned, and multialigned opportunities.

In terms of the composition of the portfolio governing body, the challenge arises that unique knowledge is required for each business opportunity. It is thus impossible for any single governance body to have the depth of understanding needed to guide each project. That is why individual projects may need oversight from a unique advisory board whose members have expertise specific to the technology or market in question and who inform the portfolio governance board of project-level issues. The composition of these boards may need to evolve as the requirements of the project change.

Appropriate Metrics and Rewards

Appropriate metrics and rewards must be in place if the innovation function and the people who staff it are to feel successful. As we've seen in earlier chapters, building breakthrough businesses can be emotionally draining. To the extent that the innovation system brings strategic renewal through transformational experiences, senior management may need to provide slack resources because of the experimental nature of that role and may hold BI to different constraints and rewards than those to which operating units are held. Both activity-based and performance-based measures may be needed since commercial success can be infrequent. Examples of appropriate innovation system metrics might include whether it has accumulated new market connections, new technical capabilities, and new partnerships or moved the firm into a new strategic domain. In one large firm, the BI system evaluates programs on how successfully the market is informed of their initiatives, whether by technical publications, conference presentations, or direct conversations with potential customer-partners. Each project must be rewarded for its learning-based progress through incubation rather than commercial success, since any given project is likely to fail. As we saw in our discussion of the learning plan, project progress may entail metrics that are not financially or market based, at least initially. It's at the portfolio level that overall contribution to the firm's growth should be considered.

Culture and Leadership That Values Breakthrough Innovation

Finally, an innovation function requires a culture and leadership who understands and values breakthrough innovation and what it takes to succeed. The ideal organizational culture and leadership team values breakthrough innovation as a key component of organizational success, acts as caretakers of the firm's future health, and understands the risks inherent in BI. While it's true that operational excellence is critical, so is innovation excellence. They're just different. This is shown in the following ways:

- Investment in strategic thinking and conversations about the future health of the firm.
- A vision of the firm's competency objectives in terms of technology platforms or market domains that senior leadership would allow to be influenced by new learning.
- The investment in technology and human capital to build and exploit those capabilities.
- A persistent drumbeat through the organization about the role of innovation as a core activity of the company. Speeches, training of general managers in leadership styles for an innovation culture, vocabulary for innovation, and career paths for innovation all serve to augment the company's belief system.

Often the reality is different from the ideal, and innovation management systems are stunted when this last system element does not align with the others.

In addition, as macroenvironmental issues change, so do senior management priorities. That is why we must all recognize the companies' capacities for innovation fluctuate. For this reason, it is critical to consider innovation as an ongoing function so that it is more deeply rooted than the current generation of management champions. We noted at the outset of this chapter that increasing numbers of senior leaders realize the need for innovation, but having been developed through the mainstream organization based on a culture of operational excellence, they may not understand what a sustainable, successful innovation function comprises. Leaders need education, support, and encouragement.

Building an innovation management system is not that difficult once its elements are clarified.

Conclusion

Established companies are experimenting with developing management systems for breakthrough innovation. It's long overdue and will help companies reduce (though not squelch) the chance factor of innovation. Systems will have these components:

- Portfolios of business opportunities tied to strategic intent rather than hidden, exception-based projects
- A cadre of experienced and appropriately trained new business creation specialists rather than a lone wolf champion and his band of followers
- Learning-oriented processes and evaluation tools
- Governance rather than hierarchy
- An identified group responsible for cultivating breakthrough innovation, with the recognition that all employees are responsible to do their part

Companies have come quite a ways in the past several years, but not fast enough. We have been privileged to be in the catbird seat, watching it unfold. We have not been studying histories, the arena of most academic works in management. Rather, we've had the rare opportunity to observe the emergence of the future and have been able to comment on its evolution.

By observing the challenges our partner companies have faced on their journeys to developing a BI capability, we have devised a set of prescriptions that can help speed the process by which companies institutionalize innovation as an ongoing, ever present business function. Each building block (discovery, incubation, and acceleration) has to be considered in its own right in terms of leadership, talent development, governance, organizational structure and location, tools, and metrics. In addition, the three building blocks must be melded as a system with feedforward and feedback loops. Finally, the whole system has to be orchestrated in line with the company's current ability to deal with innovation. Without that match, this effort will be another in the long line of defunct attempts. But as Charlie Craig, vice president of strategy at Corning told us in reflecting on Mark Newhouse's strategic growth group, "I've been here thirty years. Corning has tried to develop innovation systems many times, and I can say that this time I do believe we're onto something."

No matter where your company is in its journey toward developing an innovation capability, the concepts, frameworks, and management system we've presented should help you steer the course. Your capacity may be constrained, or it may be rich, indeed, highly demanding of innovation, but your DNA capabilities aren't developed fully. Or you may recognize in reading this book that your company understands all about discovery, but not about the other two building blocks. Any angle you start from can be a point of departure toward developing a thriving innovation function. You can't get to where you want to go unless you know where you are. We have provided a detailed map of the landscape that needs to be navigated to develop and maintain a breakthrough innovation function based on a wide variety of companies' experiences that we've tracked for a lengthy period of time. We hope you can leverage their experiences and our learning.

What's next? We fully expect to see companies developing formal career paths, hiring plans, and training for innovation. We expect to see a plethora of tools come onto the scene that help new business creation teams manage under high uncertainty. We expect culture change gurus to adopt innovation-speak. Portfolio management techniques and metrics for projects that are all high risk will burgeon. Our own research is focusing now on more deeply elaborating and testing the ideas we've presented here. One thing is for sure: we are excited to be participants in this emerging trend, and we hope that this lengthy read has not only increased your enthusiasm but offered you a path forward as well.

Appendix A

COMPANIES PARTICIPATING IN THE BREAKTHROUGH INNOVATION RESEARCH STUDY

Source: Descriptions developed from company sources and Value Line.

Phase I–Cohort I

Air Products: Supplies a variety of gases to the chemical, steel, electronics, oil, and food industries, as well as to hospitals, offices, and laboratories. Produces chemical intermediaries used in pesticides, adhesives, and coatings. Involved in the development, construction, and management of waste to energy cogeneration.

Analog Devices: Makes linear, mixed signal, and digital integrated circuits for signal processing applications. Markets to original equipment manufacturers for computers, communications, industrial, military, and other applications.

DuPont: Leader in science and technology disciplines for applications in high performance materials, electronics, safety and security, agriculture, and biotechnology. Other target markets include automotive, construction, medical, and nutrition.

General Electric: One of the largest and most diversified industrial companies in the world. Markets include infrastructure, health care, NBC, industrials, and commercial finance.

General Motors: World's second largest automotive manufacturer. Makes Chevrolet, GMC, GM diesel, and locomotive engines. Operates plants in seventeen countries.

IBM: World's largest supplier of advanced information processing technology and communications systems, services, and products.

Nortel Networks: One of the world's largest telecommunications equipment suppliers. A global leader in telephony, e-business, and wireless solutions for the Internet.

Polaroid: Filed for bankruptcy in 2001. Nearly all of its assets and the company name were acquired by Bank One. Currently, the Polaroid name is broadly licensed. Polaroid was best known for innovations in instant films and polarized glasses.

Texas Instruments: Global manufacturer of semi-conductor and electronic products. The company is the leading supplier of digital signal processors and analogue devices. Markets electrical controls and productivity solutions.

United Technologies: Operates in six major market segments including heating, ventilation and air conditioning products, elevators, jet engines, helicopters, and aircraft systems.

Phase II–Cohort II

3M: Diversified manufacturer. It sells over 50,000 products in more than 200 countries. Its seven main businesses include graphics and display technologies, health care, consumer and office products, communications, transportation, security, and protection.

Air Products: See Phase I description.

Albany International: World's largest producer of paper machine clothing. The company also manufactures high-performance doors.

Corning: Global manufacturer in the following principal market segments: display technologies, glass substrates for liquid crystal display, telecommunications, optical hardware, and environmental technologies.

DuPont: See Phase I description.

General Electric: See Phase I description.

IBM: See Phase I description.

Johnson & Johnson (Consumer Products Co.): Manufactures and sells health care products, especially baby care products, non-prescription drugs, sanitary protection products, and skin care products.

Kodak: For many years, world's leading producer of a wide range of photographic materials and equipment. In recent years, Kodak has focused on printing and digital imaging and has exited the health imaging business.

Mead-Westvaco: Global manufacturer of packaging materials, consumer and office products, and specialty chemicals.

Sealed Air Corporation: Manufacturer of specialized protective packaging products, air-cellular cushioning materials, padded mailers, and food packaging materials.

Shell Chemicals: Major world producer of commodity chemicals, including olefins, aromatics, detergent alcohols, resins, solvents, and petrochemicals. A subsidiary of Royal Dutch Shell.

Phase II–Validation Sample

Bose: Privately held U.S. company that specializes in audio equipment, car radios, automotive suspension systems, amplification systems, and home entertainment systems.

Dow-Corning: Leading silicon product manufacturer. Fifty percent owned by Corning and fifty percent owned by Dow Chemical. A global leader in silicon-based technology and innovation. Provides more than 7,000 products to world markets.

Guidant: Recently acquired by Boston Scientific. Designs and manufactures pacemakers, defibrillators, stents, vascular intervention products, and innovative cardiovascular products.

Hewlett-Packard: Provides computing, printing, and imaging solutions and services to consumers and businesses. Main markets include personal supplies, enterprises storage and servers, HP services, financing, and software.

Intel: Leading manufacturer of integrated circuits for personal computers, communications, industrial automation, and military equipment. Main products are microprocessors and memory chips.

Procter & Gamble: Makes and markets detergents, soaps, foods, paper, and industrial products. Brands include Head & Shoulders, Olay, Downy, Tide, Bounty, Folgers, Iams, Pringles, Gillette, and Duracell.

PPG: Global manufacturer of glass products, coatings, resins and industrial specialty coatings, automotive and industrial coatings, fiberglass, float glass, and caustic soda.

Rohm & Haas: Specialty chemicals producer in acrylic technologies. Products include coatings, adhesives, plastics, herbicides, biocides, fungicides, exchange resins, and electronic materials.

Xerox: Develops, manufactures, markets, finances, and services copiers, laser printers, and document publishing equipment.

Appendix B

ASSESSING YOUR FIRM'S BREAKTHROUGH INNOVATION COMPETENCY

We constructed a series of questions based on research into the attributes of breakthrough innovation. This research identified seventeen areas of breakthrough innovation competencies that together comprise a business innovation management system. This questionnaire will help you and your colleagues assess the strengths and needs for improvement of your firm's breakthrough innovation system and skills.

The following series of statements are related to the competencies discussed in each chapter, and each set of statements results in an average for that competence. For each competence, assess the current level of competency in your organization for that element and the desired level that could reasonably be attained in your company in eighteen months.

In the box beside each statement, list the number of the appropriate response for each item using this system:

1 = I strongly disagree with this statement.

2 = I disagree somewhat with this statement.

3 = I neither agree nor disagree with this statement.

4 = I agree somewhat with this statement.

5 = I strongly agree with this statement.

For each section, average your results. Place your answers in the summary table at the end of the survey and plot them on the graph that follows to get a visual representation of your company's innovation maturity level. In some of the questions, by "senior management," we mean those managers who influence the innovation activities in your firm. If your firm operates in fairly autonomous divisions, use your division as your reference organization.

Chapter Two

2.1 Organizational Awareness of Breakthrough Innovation

		As Is	Desired
1.	Our firm recognizes that breakthrough ideas must be managed differently from conventional new product development projects.		
2.	New employees are made aware that different types of innovation projects exist in our organization. Some have real breakthrough potential, and others are incremental or programmatic improvements to current product lines.		
3.	In our firm, we treat most new product development projects and new technology projects differently.		
4.	In our firm, the importance of breakthrough innovation for our long-term success is well understood.		
5.	Breakthrough innovation is a concept that is well understood in our firm.		
6.	Our firm has initiatives oriented toward breakthrough innovation at this time.		
	Total		

2.2 The Role of Senior Management in Breakthrough Innovation

		As Is	Desired
1.	Senior management thinks about the future of technology trends that are relevant to us and can articulate a compelling vision.		
2.	Senior management drives the generation of breakthrough innovation ideas in our firm.		
3.	In the past five years, senior management has increased its emphasis on the importance of breakthrough innovation.		
4.	Senior management sets the context, strategic direction, and framing for breakthrough innovation.		
5.	Senior management is committed to breakthrough innovation in our firm.		
6.	Senior management works to develop an innovation supportive organizational culture.		
	Total		

2.3 Role of Research and Development in Stimulating Breakthrough Innovation

	As Is	Desired
1. Scientists have mechanisms for staying aware of new technical developments outside our firm's current technical core.		
2. In our firm, scientists are encouraged to explore new ways for combining technologies that may lead to novel discoveries.		
3. Scientists work with external partners to explore new ideas.		
4. Scientists are encouraged to proceed with technical development in our firm even though they may not have a clear understanding of the fundamental underlying science.		
5. Our firm allows scientists a portion of their time to work on new ideas that may yield breakthrough innovations.		
6. Scientists are aware of how our firm's strategic intent influences the pursuit of a particular technology.		
Total		

Chapter Three

3.1 Generating Breakthrough Innovation Ideas

	As Is	*Desired*
1. There is a group of people in our organization whose responsibility it is to receive ideas and work with the idea generators to consider the commercial opportunity of the concept.		
2. We have a steady flow of ideas that are potential breakthrough innovations.		
3. Even though an idea does not have business unit support from the beginning, it will still be pursued in our organization.		
4. We encourage all employees in our firm to think creatively about future business opportunities.		
5. In our firm, there is an electronic system for collecting breakthrough innovation ideas.		
6. We encourage people to come forward with ideas, even if they have only a vague idea of the potential market applications for the idea.		
Total		

3.2 Recognizing Breakthrough Innovation Opportunities

	As Is	Desired
1. We find ways to make sure our scientists get connected to outside sources of new knowledge.		
2. In our firm, if someone has an idea but cannot find an immediate champion for it, there is still a place for that person to take that idea.		
3. We have people in our organization who fit the description of hunters: they go around sniffing out potential innovation opportunities.		
4. We have a clear system for evaluating the breakthrough innovation potential of submitted ideas.		
5. We have programs in place to build and cultivate informal networks within the organization for providing feedback on the technical feasibility of ideas.		
6. In our organization, people who come up with innovation ideas are assisted in developing a rationale for the potential of their ideas.		
Total		

3.3 Evaluating Breakthrough Innovation
Opportunities

	As Is	*Desired*

1. In our firm, we apply different evaluation criteria for breakthrough innovation opportunities than we use to evaluate incremental innovations.

2. We bring in outside expertise to help us evaluate breakthrough opportunities that are not part of our current competency base.

3. Breakthrough innovation projects in our organization can still be pursued even if they do not fit into our current customers' needs.

4. We have people who evaluate breakthrough innovation opportunities who do not come out of the mainstream lines of business.

5. In order to get approval to pursue a potentially breakthrough opportunity in our firm, it is not necessary that the final market opportunity or the eventual business be clear from the outset.

6. In order to get approval to pursue a potentially breakthrough opportunity in our firm, a clear business case and project plan does not need to be developed first.

   Total

Chapter Four

4.1 Managing Breakthrough Innovation Projects from Inception to Transition

		As Is	*Desired*
1.	Our firm understands that projects cannot all be managed with the same sets of tools and techniques.		
2.	People seek to manage breakthrough innovation projects in our organization even though those projects are risky.		
3.	We acknowledge that a project manager's required expertise for breakthrough innovation must be more about anticipating and monitoring uncertainties than about planning, controlling, and delegating.		
4.	The people we assign as breakthrough innovation project managers thrive under conditions of high uncertainty.		
5.	Our project managers are skilled at getting organizational buy-in for their projects.		
6.	When we put together a team for a breakthrough innovation project, we make sure they understand the nature of the breakthrough innovation life cycle, so they'll know what to expect.		
	Total		

4.2 Learning About Markets for
Breakthrough Innovation

	As Is	Desired
1. To learn about unfamiliar markets, we rely on lead users and prototypes rather than customer surveys or focus groups.		
2. We investigate initial markets with applications that are easiest to enter rather than the biggest potential market.		
3. We don't always rely on our current customers to give us feedback about our new breakthrough ideas.		
4. Members of breakthrough innovation teams in our organization try to generate a list of as many applications for the innovation as possible before choosing which to pursue.		
5. We have marketing personnel in our organization who have expertise in exploring unfamiliar markets for the purposes of learning about those markets.		
6. Our firm has skilled business development staff to help learn about unfamiliar markets.		
Total		

4.3 Developing the Business Model for Breakthrough Innovation

		As Is	Desired
1.	We understand our eventual place in the value chain that we can stake out.		
2.	As we develop breakthrough innovation opportunities, we constantly ask ourselves what the eventual business model will be.		
3.	We devote resources to understanding the entire value chain that our innovation will require.		
4.	The issue of how to make money is usually surfaced with a project team while they are doing early technical development work.		
5.	Exploring partnerships with other members of the value chain is an important part of the project team's mandate for breakthrough innovation projects.		
6.	Project teams are expected to learn about and manage early market development issues for their breakthrough innovation projects.		
	Total		

4.4 Acquiring Funding for Breakthrough Innovations

	As Is	*Desired*
1. Our firm funds breakthrough innovation project development differently than they fund incremental innovation projects.		
2. To augment internal funding, our firm assists project teams in finding external funding.		
3. Our funding process for projects is staged, based on their demonstrated progress.		
4. Our firm uses a portfolio approach for evaluating and funding innovation projects.		
5. We have an organizational process in place for funding breakthrough innovation projects that is separate from other types of innovation projects.		
6. We have a corporate-level fund in place for funding breakthrough innovation projects.		
Total		

4.5 Acquiring Breakthrough Innovation People

	As Is	Desired
1. People who have entrepreneurial mind-sets are highly valued in our firm.		
2. In our firm, we do not punish people who participate in failed breakthrough innovation projects.		
3. In our firm, participation in breakthrough innovation activities is seen as a broadening, career-enhancing experience.		
4. Career development for our high-potential scientists and technical people includes mechanisms for broadening their experience base into other areas of science and business.		
5. There is a career development path for people who have entrepreneurial mind-sets.		
6. Our firm looks explicitly for people who have entrepreneurial mind-sets. We do not necessarily expect them to emerge from our existing personnel.		
Total		

Chapter Five

5.1 Develop an Acceleration Competency

	As Is	Desired
1. We have an identified and formal acceleration leader or organization along with top management support.		
2. There are resources available to support fast growth of breakthrough businesses in acceleration.		
3. There is a clear way to move breakthrough businesses to operations.		
4. We have recruited and trained a set of acceleration managers.		
5. We have developed rich internal and external networks to build a critical mass of our breakthrough businesses.		
6. We have ensured a strategic fit with our company, or we have redirected company strategy to fit an identified opportunity.		
Total		

5.2 Acceleration Requirements

	As Is	Desired

1. We view the achievement of acceptable manufacturing yields as a key responsibility of the acceleration function.

2. To make the transition to their ultimate organizational home, breakthrough businesses must be able to demonstrate a pathway to profitability.

3. Once an accelerating business can provide comfortably predictable sales forecasts and demonstrate repeat customers, we move it to operations.

4. We have a process for moving our businesses out of acceleration and into operations at the right time for business unit acceptance.

5. Businesses in acceleration have an identified and willing organizational home in an existing business unit or a new one by the time they're ready for the transition.

6. We communicate acceleration status and progress for each business in acceleration to other acceleration businesses and to project managers in incubation.

Total

5.3 Drive New Business to Where It Can Stand on Its Own

		As Is	*Desired*
1.	We ensure our accelerating businesses have a sales and marketing infrastructure.		
2.	We ensure our accelerating businesses have an operations infrastructure.		
3.	We devote appropriate attention to building the general management team for our accelerating businesses.		
4.	Each accelerating business has a senior-level management advisory board to provide support for it within the company.		
5.	The metrics for measuring progress for each business in acceleration are clear.		
6.	Management in the organizational home participates in the oversight for each accelerating business to prepare it for transition to the operating home.		
	Total		

Chapter Six

6.1 Managing Innovation System Resources

	As Is	Desired

1. Breakthrough innovation projects have sufficient resource commitment by our firm.

2. In our firm, we have decision rules for removing resources and killing projects.

3. Our firm uses an investment approach to innovation that is similar to a venture capital mind-set.

4. There is a system in our firm to provide an appropriate level of resources at whatever level of maturity the project is at.

5. The decision criteria for allocating resources to breakthrough innovation projects in our firm are clear.

6. The process for accessing and leveraging corporate resources (funding, personnel, capital equipment, and other organizational resources) for breakthrough innovation projects in our firm is clear.

Total

6.2 Integrating Innovation System Interfaces

		As Is	*Desired*
1.	The innovation processes in our firm are considered to be a system of interacting functions or responsibilities.		
2.	There is clarity of purpose for each part of the system in our firm.		
3.	Our firm has developed processes to transfer project responsibilities for breakthrough innovation projects as they make the transition from one point in the system to the next.		
4.	Each part of the firm's innovation system interacts with each other regularly.		
5.	In our firm, organizational members are aware of their own roles in the innovation system.		
6.	Members of the firm are aware of the roles and responsibilities of each unit in the innovation system.		
	Total		

6.3 Portfolio-Level Thinking

		As Is	*Desired*
1.	We think about funding of innovation projects as a portfolio of investments.		
2.	We diversify our innovation investments along dimensions of risk.		
3.	Senior managers make decisions about the innovation investments as part of their strategy discussions.		
4.	Senior management spends time and attention on our breakthrough innovation portfolio.		
5.	Our firm has a well-articulated and broadly communicated strategy concerning investing in innovation projects.		
6.	Our firm evaluates return on invested capital in innovation projects for our entire set of innovation investments.		
	Total		

6.4 Portfolio Management

	As Is	*Desired*
1. Overall, our company has an adequate number of major innovation projects in its portfolio.		
2. Each of the projects in the portfolio constitutes a portion of a balanced portfolio of investments, given each project's stage of maturity.		
3. Our major innovation projects form the basis of a continuing pipeline of potentially significant businesses.		
4. The team that manages the portfolio exhibits an "options" mentality as it pertains to individual project selection.		
5. The team responsible for the portfolio of major innovation projects exhibits a strong experimental orientation.		
6. The team responsible for the portfolio of major innovation projects is populated with managers who have previously struggled with the entire innovation process from discovery through commercialization.		
Total		

6.5 Breakthrough Innovation Organization Systems

	As Is	Desired

We have an organizational system that:

1. Acts as a known home for anyone with a breakthrough idea to come to.

2. Helps develop breakthrough ideas into valuable business concepts.

3. Helps build project teams for promising business concepts.

4. Evaluates and oversees the portfolio of breakthrough innovation projects.

5. Assembles appropriate advisory boards for each potential opportunity.

6. Acts as a repository of skills, expertise, and information on managing breakthrough innovation.

7. Has extensive networks within the organization to draw on for necessary expertise.

8. Can link project initiators with appropriate sources of funding, both within and external to the organization.

9. Facilitates the interfaces among various groups that participate in the innovation system.

10. Considers our innovation project investments as a portfolio of investments.

11. Consists of a balanced pipeline of innovation projects.

    Total

Chapter Seven

7.1 Orchestration in Our Organization

		As Is	Desired
1.	Our organization realizes the need for integration among our discovery, incubation, and acceleration activities.		
2.	We understand our organization's capacity for doing innovation at any point in time.		
3.	Orchestration is appropriate to our organization's innovation capacity.		
4.	Our organization understands orchestration activities.		
5.	Our organization has an overall view of all our innovation activities.		
6.	Our organization audits the interactions among our innovation activities.		
	Total		

7.2 Orchestrators in Our Organization

	As Its	Desired
1. We have an orchestration role in our organization.		
2. The person who plays this role in our organization has influence.		
3. The person who plays the orchestration role in our organization understands the strategic intent of our company.		
4. The person who plays the orchestration role in our organization is recognized as important in our organization.		
5. The person who plays the orchestration role in our organization works to maintain balance among the parts of our innovation system.		
6. The person who plays the orchestration role in our organization works both inside the organization and with outside constituencies for ensuring the success of the DNA system.		
Total		

Note: Although we have used "person" in this part, there can be more than one person in the orchestration role.

Competency Grand Score and Analysis

For each competency, indicate in the table that follows:

- The extent to which your firm is strong or requires development, based on the score you received relative to the desired score—that is, the extent to which there are gaps between current and desired conditions.

- The extent to which you believe each item is important for the establishment of breakthrough innovation in your firm, given your company's culture and strategic priorities: high (H), medium (M), or low (L) importance.

- Areas that have large discrepancy scores or where the As Is score is lower than you would have liked indicate particular areas of concern.

Breakthrough Innovation Competency Areas	Maximum Score	As Is Score	Desired Score	Difference	Importance (H, M, L)
2.1 Organizational Awareness of Breakthrough Innovation	30				
2.2 The Role of Senior Management in Breakthrough Innovation	30				
2.3 Role of Research and Development in Stimulating Breakthrough Innovation	30				
3.1 Generating Breakthrough Innovation Ideas	30				
3.2 Recognizing Breakthrough Innovation Opportunities	30				
3.3 Evaluating Breakthrough Innovation Opportunities	30				
4.1 Managing Breakthrough Innovation Projects from Inception to Transition	30				
4.2 Learning About Markets for Breakthrough Innovation	30				
4.3 Developing the Business Model for Breakthrough Innovation	30				
4.4 Acquiring Funding for Breakthrough Innovations	30				

Breakthrough Innovation Competency Areas	Maximum Score	As Is Score	Desired Score	Difference	Importance (H, M, L)
4.5 Acquiring Breakthrough Innovation People	30				
5.1 Develop an Acceleration Competency	30				
5.2 Acceleration Project Requirements	30				
5.3 Drive New Business to Where It Can Stand on Its Own	30				
6.1 Managing Innovation System Resources	30				
6.2 Integrating Innovation System Interfaces	30				
6.3 Portfolio-Level Thinking	30				
6.4 Portfolio Management	30				
6.5 Breakthrough Innovation Organization Systems	55				
7.1 Orchestration in Our Organization	30				
7.2 Orchestrators in Our Organization	30				
Total					
Total possible	655				

Now plot your scores on the graph below to see a visual representation of your BI competency analysis. First plot your "as is" score in one color, then plot your "desired" score in a second color. Go to work first on the areas where the gap is the greatest and that you've indicated have an importance of medium or high levels to your company.

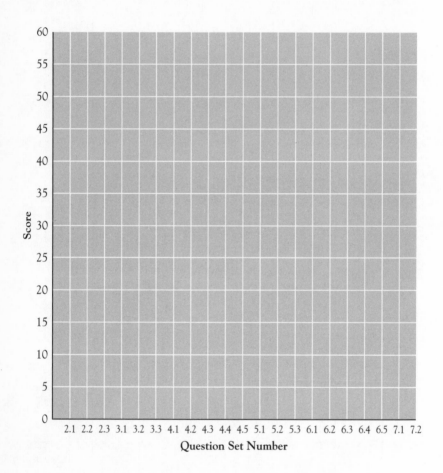

Notes

Preface

1. Leifer, R., McDermott, C., O'Connor, G. C., Peters, L., Rice, M., & Veryzer, R. (2000). *Radical innovation: How mature firms can outsmart upstarts*. Boston: Harvard Business School Press.
2. The research team members, to whom we give grateful acknowledgment, were Andrew Corbett, Richard DeMartino, P. J. Guinan, Donna Kelley, Heidi Neck, T. Ravichandran, and Dan Robeson. We also acknowledge the cochairs of the IRI subcommittee on Radical Innovation: Alan Ayers of Energizer, Dave McKeough of Pittsburgh Plate and Glass, Ian Elsum of CSIRO, Ted Farrington of J&J Consumer Products, and Steve May of Exxon-Mobil.

Introduction

1. Pinchot, G. (1985). *Intrapreneuring: Why you don't have to leave the corporation to become an entrepreneur*. New York: Harper-Collins.
2. Leifer, R., McDermott, C., O'Connor, G. C., Peters, L., Rice, M., & Veryzer, R. (2000). *Radical innovation: How mature firms can outsmart upstarts*. Boston: Harvard Business School Press; O'Connor, G. C., & Rice, M. P. (2001). Opportunity recognition and breakthrough innovation in large established firms. *California Management Review, 43*(2), 95–116.

3. Single, A. W., & Spurgeon, W. M. (1996). Creating and commercializing innovation inside a skunk works. *Research-Technology Management, 39*(1), 38–41.

4. Ginsberg, A. (2001). Blue chip entrepreneurs: Corporate venturing and creative destruction. Retrieved Sept. 30, 2007, from http://w4.stern.nyu.edu/emplibrary/Ginsberg_Pubs.pdf.

5. Ginsberg. (2001).

6. Ginsberg. (2001).

7. Ginsberg. (2001).

8. In a study comparing over thirty thousand investments in start-ups over a fifteen-year period, Professors Paul Gompers and Josh Lerner found that corporate venture investments appear to be at least as successful (using such measures as the probability of a portfolio firm going public) as traditional venture capital firms. In another study analyzing over three hundred venture capital–backed, information technology initial public offerings in 1998–1999, researchers Marku Maula and Gordon Murray demonstrated that emerging technology companies performed better with corporate equity investments than with traditional venture capital investors. The financial involvement of global Fortune 500 information companies was directly associated with higher first-day valuations. In both studies, success is correlated with the strategic fit between the corporate parent or corporate investor and the venture. See Gompers, P., & Lerner, J. (2001). The venture capital revolution. *Journal of Economic Perspectives, 15*(2), 145–168; Maula, M.V.J., Autio, E., Murray, G. (2005). Corporate venture capitalists and independent venture capitalists: What do they know, who do they know, and should entrepreneurs care? *Venture Capital: An International Journal of Entrepreneurial Finance, 7*(1), 3–19.

9. Robert Burgelman's studies of internal ventures within a new venture division of one firm are informative about the vulnerability of this organizational form for BI or new business creation. His (1983) findings suggest strongly that the engine of corporate entrepreneurship or ability to foster successful internal

corporate venturing resides in the autonomous strategic initiative of individuals at the operational levels in the organization. Following that insight, he comes up with some design alternatives based on varying administrative and operational linkages for individual corporate ventures (Burgelman, 1984). While he looks at internal corporate ventures in a new venture division of one company, he does not spend much time on the processes or procedures followed in that division for managing or building a portfolio of ICVs. In a later article, he does describe the new venture division of his study company and compares it to other operating divisions of that company, not new venture divisions or Breakthrough Innovation systems of other companies. His analysis sheds further light on the unstable position of the new venture divisions (NVD) in the large divisionalized firm as described by Fast (1979).

He examines why this instability may be necessary from the corporate point of view and suggests that corporate managements have not yet learned to use the NVD well. He points out that there seems to be a tendency to use the NVD more for control purposes as opposed to a tool for managing the entrepreneurial part of the strategic process. The NVD according to Burgelman is suited well for ambiguous situations, but he speculates that in other situations, different types of new venture arrangements may be more appropriate. He then stresses the need for more research to specify more clearly the conditions under which such alternative arrangements could be appropriate.

See Burgelman, R. (1986). Managing corporate entrepreneurship. In M. Horwitch (Ed.), *Technology in the modern corporation*. New York: Pergamon Press; Burgelman, R. A., & Sayles, L. R. (1986). *Inside corporate innovation*. New York: Free Press; Burgelman, R. A. (1983). A process model of internal corporate venturing in the diversified major firm. *Administrative Science Quarterly, 28,* 223–244; Burgelman, R. A. (1984). Designs for corporate entrepreneurship in established firms. *California Management Review, 26*(3), 154–166; Fast, N. D.

(1979). The future of industrial new venture departments. *Industrial Marketing Management, 8,* 264–273.

10. Boyle, M. (2006, November 27). IBM goes shopping. *Fortune, 154*(11), S2, 77.

11. Leifer, R. G., Kasthurirangan, D. R., & Mahsud, R. (2006). *R&D investment and innovativeness: Their contributions to organizational success.* Paper presented at the Strategic Management Society Annual Meeting, Vienna, Austria.

Chapter One

1. Leifer, R., McDermott, C., O'Connor, G. C., Peters, L., Rice, M., & Veryzer, R. (2000). *Radical innovation: How mature firms can outsmart upstarts.* Boston: Harvard Business School Press.

2. See, for example, Cho, H.-J., & Pucik, V. (2005). Relationship between innovativeness, quality, growth, profitability and market value. *Strategic Management Journal, 26,* 555–575; Zahra, S. A. (1991). Predictors and financial outcomes of corporate entrepreneurship: An exploratory study. *Journal of Business Venturing, 6*(4), 259–285; Sorescu, A. B., Chandy, R. K., & Prabhu, J. C. (2003). Sources and financial consequences of radical innovation: Insights from pharmaceuticals. *Journal of Marketing, 67,* 82–102.

3. Leifer et al. (2000).

4. Morone, J. G. (1993). *Winning in high-tech markets: The role of general management.* Boston: Harvard Business School Press; Govindarajan, V., & Trimble, C. (2005). *Ten rules for strategic innovators: From idea to execution.* Boston: Harvard Business School Press; Kim, W. C., & Mauborgne, R. (2005). *Blue ocean strategy: How to create uncontested market space and make the competition irrelevant.* Boston: Harvard Business School Press.

5. Bower, J. L., & Christensen, C. M. (1995, January–February). Disruptive technologies: Catching the wave. *Harvard Business Review,* 43–53.

6. Sorescu et al. (2003).
7. Govindarajan & Trimble. (2005); Kim & Mauborgne. (2005); Tucker, R. B. (2002). *Driving growth through innovation: How leading firms are transforming their future*. San Francisco: Berrett-Koehler.
8. In the earlier phase of our research, we defined a radical innovation as one that management believed had the potential to offer new-to-the-world performance features, significant (five to ten times) improvement in known features, or significant (thirty to fifty percent) reduction in cost. This definition was arrived at in conjunction with the membership of the Industrial Research Institute. It helped identify particular projects for inclusion in our phase I study, but we find that these order-of-magnitude measures varied dramatically by industry, and so made our definition more general in phase II.
9. Fast, N. D. (1978). New venture departments: Organizing for innovation. *Industrial Marketing Management, 7*(2), 77; Lerner, J., & Gompers, P. A., *The venture capital cycle*. Cambridge, MA: MIT Press, cited in Garvin, D. A., & Levesque, L. C. (2006). Meeting the challenge of corporate entrepreneurship. *Harvard Business Review, 84*(10), 102.

Chapter Two

1. Much of this GE mini-case was developed by Mahesh Shankar, an M.B.A. student at RPI, in spring 2006.
2. GE Finds its inner Edison. (2003, October). *Technology Review*, 46–50.
3. Interviews with Steve Arthur and Tom McElhenny, May 1, 2002.
4. Interview with Scott Donnelly, May 7, 2002.
5. Brady, D. (2005, March 28). The Immelt revolution: He's turning GE's culture upside down, demanding far more risk and innovation. *BusinessWeek*.

6. Much of this Diversified Industries mini-case was developed by Kyewon Maeng, an M.B.A. student at RPI, in spring 2006.

Chapter Three

1. National Science Board. (2006). *Science and engineering indicators 2006*. (Vol. 1, NSB 06–01, chap. 4). Arlington, VA: National Science Board, National Science Foundation. Retrieved Sept. 30, 2007, from http://www.nsf.gov/statistics/seind04/toc.htm. Print copies available at paperpubs@nsf.gov or (703) 292–7827.
2. Armbrecht, F.M.R. Jr. (2004). R&D and innovation in industry. In American Association for the Advancement of Science, *Report 28: R&D FY 2004* (pp. 31–37). Washington, DC: Author.
3. National Science Board. (2006).
4. Chesbrough, H. W. (2003). *Open innovation: The new imperative for creating and profiting from technology*. Boston: Harvard Business School Press.
5. The 2005 U.S. Patent and Trademark Office Report indicates IBM has held this position for thirteen consecutive years. Retrieved Sept. 30, 2007, from http://news.zdnet.co.uk/itmanagement/0,1000000308,39246274,00.htm.
6. Baghai, M., Coley, S., & White, D. (1999). *The alchemy of growth: Practical insights for building the enduring enterprise*. New York: Perseus.
7. Graham, M.B.W. & Shuldiner, A. T. (2001). *Corning and the craft of innovation*. New York: Oxford University Press. Also see http://en.wikipedia.org/wiki/Corning_Inc.
8. The figures on the technical community size were provided by Charlie Craig, vice president for science and technology at Corning, in November 2006.

Chapter Four

1. Sellers, P. (2006, November 13). The net's next phase. *Fortune*, *154*(10), 71–72.

2. Lynn, G. S., Morone, J. G., & Paulson, A. S. (1996). Marketing and discontinuous innovation: The probe and learn process. *California Management Review, 38*(3), 8–37; Lynn, G. (1993). *Understanding products and markets for radical innovation* (pp. 15–62). Unpublished doctoral dissertation, Rensselaer Polytechnic Institute, Troy, New York.

3. Much of this section is adapted from Rice, M. P., O'Connor, G. C., & Pierantozzi, R. (2008, Winter). Implementing a learning plan to counter project uncertainty. *Sloan Management Review.*

4. Lynn, Morone, & Paulson (1996); Govindarajan, V., & Trimble, C. (2005). Ten rules for strategic innovators: From idea to execution, (chap. 6). Boston: Harvard Business School Press; McGrath, R. G. & Macmillan, I. C. *Market busters: 40 strategic moves that drive exceptional business growth* (p. 197). Boston: Harvard Business School Press.

5. Hindo, B. (2007, June). At 3M, a struggle between efficiency and creativity. *Business Week Innovation* (insert), 8–15.

Chapter Five

1. Kodak turned its first profit in two years in the first quarter of 2007, the fourth year of a four-year turnaround plan. Interestingly, much of the quarter's profits came from settlements of intellectual property suits and licensing fees: technology Kodak had that others were commercializing. Bulkeley, W. M. (2007, February 1). Kodak takes hit in film and digital. *Wall Street Journal*, B3.

2. For a useful tool to help with the transition of accelerating businesses to operating units, see O'Connor, G. C., Hendricks, R., & Rice, M. P. (2002, November–December). Assessing transition readiness for radical innovations. *Research-Technology Management*, pp. 50–56.

3. Leifer, R., McDermott, C., O'Connor, G. C., Peters, L., Rice, M., & Veryzer, R. (2000). *Radical innovation: How mature firms can outsmart upstarts*. Boston: Harvard Business School Press.

4. Moore, G. A. (1991). *Crossing the chasm*. New York: HarperCollins.
5. Norling, P., & Statz, Robert J. (1998, May–June). How discontinuous innovation really happens. *Research-Technology Management*, 41–44.
6. Dyer, G., & Firn, D. (2003, October 11). GE puts faith in personalised medicine with purchase of UK diagnostics leader. *Financial Times*, p. 5.

Chapter Six

1. O'Connor, G. C., & DeMartino, R. (2006). Organizing for radical innovation: An exploratory study of the structural aspects of RI management systems in large established firms. *Journal of Product Innovation Management, 23*, 475–497.
2. Chesbrough, H. (2003). *Open innovation: The new imperative for creating and profiting from technology*. Boston: Harvard Business School Press; Chesbrough, H. (2006). *Open business models: How to thrive in the new innovation landscape*. Boston: Harvard Business School Press.
3. Stevens, G. A., & Burley, J. (2003, March/April). Piloting the rocket of radical innovation. *Research Technology Management, 46*(2), 16–25.
4. We're not the only ones. Our participating companies echoed this expectation, and other authors have as well. See, for example, Chesbrough, H. W. (2000). Designing corporate ventures in the shadow of private venture capital. *California Management Review, 42*(3), 31–49.
5. For a useful tool to aid in businesses' transition to operations, developed in conjunction with the Industrial Research Institute, see O'Connor, G. C., Hyland, J., & Rice, R. (2006). Bringing radical innovations successfully to market: Bridging the transition from R&D to operations. In Woodland, B. C., & Miller, C. W. (Eds.). (2004). *Creating the customer connection: PDMA toolbook II for new product development*.

New York: Wiley. See also O'Connor, G. C., Hendricks, R., & Rice, M. P. (2002, November–December). Assessing transition readiness for radical innovations. *Research-Technology Management*, 50–56.

6. Colvin, G. (2005, September 19). The bionic manager. *Fortune*, 89–92.

7. Corporate R&D scorecard. (2005, September). *Technology Review*, 108(9), 56–61; Leifer, R., Kasthurirangan, G., Robeson, D., & Mahsud, R. (2006). *R&D investment, and innovativeness: Their contributions to organizational success*. Paper presented at the Strategic Management Society Annual Meeting, Vienna, Austria.

8. Rae, J. (2006, April 28). Closing the gap on innovation metrics. *Business Week Online*. www.businessweek.com. Retrieved Sept. 30, 2007, from http://www.businessweek.com/innovate/content/apr2006/id20060428_377270.htm?campaign_id=search.

Chapter Seven

1. Norling, P., & Statz, R. J. (1998, May–June). How discontinuous innovation really happens. *Research Technology Management*, 41–44.

2. For more details on the issues associated with transition management, see Rice, M. P., Leifer, R., & O'Connor, G. C. (2002). Commercializing discontinuous innovations: Bridging the gap from discontinuous innovation project to operations. *IEEE Transactions on Engineering Management, 49*, 330–340.

3. We refer to journals such as the *Academy of Management Journal, Academy of Management Review, Journal of Product Innovation Management, Management Science*, and many others.

4. Burgelman, R. A. (1983). A process model of internal corporate venturing in the diversified major firm. *Administrative Science Quarterly, 28*, 223–244.

Chapter Eight

1. Campbell, A., Birkenshaw, J., Morrison, A., & van Basten Batenburg, R. (2003, Fall). The future of corporate venturing. *Sloan Management Review*, 30–37.

2. See, for example, Hee-Jae, C., & Pucik, V. (2005). Relationship between innovativeness, quality, growth, profitability and market value. *Strategic Management Journal*, *26*, 555–575; Zahra, S. A. (1991). Predictors and financial outcomes of corporate entrepreneurship—an exploratory-study. *Journal of Business Venturing*, *6*(4), 259–285; Sorescu, A. B., Chandy, R. K., & Prabhu, J. C. (2003). Sources and financial consequences of radical innovation: Insights from pharmaceuticals. *Journal of Marketing*, *67*, 82–102.

3. Kim, W. C., & Mauborgne, R. (2005). *Blue ocean strategy: How to create uncontested market space and make the competition irrelevant*. Boston: Harvard Business School Press.

Acknowledgments

We would like to acknowledge the help and support of a number of people who provided the opportunity for us to engage in this very exciting research program, stimulated and challenged our thinking, and encouraged us to write this book.

First, we must express our appreciation for the role that the Industrial Research Institute (IRI) played. Staff at the Institute were a constant support to us as a listening post, a critical reviewer of our work, and a partner in every sense. They adopted our research agenda and guided it to ensure its relevance and robustness. Ross Armbrecht, president of IRI while we conducted our research program, provided support, help, and commitment. The IRI's Research on Research Committee, which established the Radical Innovation (RI) subcommittee, met three times each year to work with us, hear our findings and help us interpret them, complete numerous surveys, and remind us of the importance of the work. The RI subcommittee was co-chaired by a running list of executives who gave of their limited time and talented insight: Alan Ayers of Energizer, Dave McKeough of Pittsburgh Plate and Glass, Dave Austgen of Shell Chemicals, Ted Farrington of Johnson & Johnson Consumer Products, Ethan Simon of Rohm & Haas, Ian Elsum of CSIRO, and Steve May of Exxon-Mobil. If we had not had that subcommittee, those co-chairs, and the loyal members of the IRI to turn to, we simply could not have convinced our participating companies to partner with us on this four-year journey.

Second, our partner companies must be acknowledged. The innovation engines in each of these companies boiled down to a few people who continuously analyzed and worked to improve their organization's innovation efforts, and they spent the time and effort to be open and honest with us during each six-month check-up. We learned about them and we learned with them. Although we run the risk of excluding important people that we encountered in each of the companies, we'd like to gratefully acknowledge the input of the following participants with whom we spoke on multiple occasions: Paul Guehler, Jay Ihlenfeld, Roger Lacey, Michelle Nelson, and Debra Wilfong at 3M; Ed Hahn and Charlie Kramer at Albany International; Miles Drake, Nancy Easterbrook, John Marsland, Ron Pierantozzi, Chris Sutton, and Larry Thomas at Air Products; Charlie Craig, Joe Miller, Deb Mills, and Mark Newhouse at Corning; Larry Berger, Dick Bingham, Michael Blaustein, and John Hillenbrand at DuPont; Scott Donnelly, Howard Goldberg, Michael Idelchik, Nancy Martin, and Rosanna Stokes at GE; Letina Connelly, Mike Giersch, Bruce Harreld, and Scott Penberthy at IBM; Margaret Aleles, Ted Farrington, Neal Mattheson, and Kurt Schilling at Johnson & Johnson Consumer Products; Ed Covannon, Gary Einhaus, Greg Faust, Larry Henderson, Jose Mir, Rich Notargiacomo, Nancy Sousa, and Dana Wolcott at Kodak; Don Armagnac, Laura Pingel, Fred Renk, Rick Spedden at Mead-Westvaco; Ron Cotterman, Bill Hickey, and Larry Pillotte at Sealed Air Corporation; Dave Austgen, Howard Fong, Joe Machado, Brendan Murray, Tom Semple, and Bob Tait at Shell Chemicals.

As well, our validation group of companies—who journeyed to see us at Rensselaer Polytechnic Institute (RPI) four times over the course of our study program to share their experiences and provide feedback on our interpretation of the data—were a rich treasure trove of insight, patience, and learning. We gratefully acknowledge and thank Joe Baumann of Procter & Gamble; Scott Boyce, Sanjay Chaturvedi, and Ethan Simon of Rohm & Haas; Dave McKeough and Alan Wang of Pittsburgh Plate and Glass; Tom Workentine and Greg Zank of Dow-Corning; John

Capek and Sami Hamade of Guidant; Angela Biever and Ohmid Moghadam of Intel; Santokh Bandesha and Debbie Wickham of Xerox; Jim Cook and Tim Weber of Hewlett-Packard; and Brian Mulcahy and Dave Thomas of Bose. Special thanks to Shreefal Mehta—the executive director of the Radical Innovation Research Program during our study—who organized and ran the series of meetings with our validation companies. He set the standard for high levels of debate, well-managed communications, and perfectly executed meetings. As a result, all participants gained great insight, provided wonderful contributions, and developed rich networks.

We thank Linda Arnoldussen of the University of Alberta School of Business Executive Education and Lifelong Learning Division for her insight in suggesting the key term *DNA of breakthrough innovation* to describe Discovery, iNcubation, and Acceleration. Other supporters included Joanne Hyland, former vice president of the Business Ventures Group at Nortel Networks and now head of the Radical Innovation group, and Michael Wolff, editor of the IRI's publication *Research-Technology Management*. In addition, the NSF provided partial support for this research through RPI's Nanoscale Engineering Research Center.

We especially thank the former and current deans of Lally School of Management and Technology at RPI—Denis Simon and David Gautschi—for their ongoing support and guidance in managing our complex research program. Jeannette Bines and Jill Keyes provided constant, reliable, and much-needed administrative help; we thank you both.

Finally, we gratefully acknowledge our research team, who helped collect the data, sat through our summer data analysis workshops, formulated many rich ideas, and wrote numerous papers based on what we learned: Andrew Corbett, T. Ravichandran, and Dan Robeson at RPI; Richard DeMartino at Rochester Institute of Technology; and P.J. Guinan, Donna Kelly, and Heidi Neck at Babson College. Their contributions to scholarship, to the research enterprise, and to helping companies find the path to building an innovation capability were immeasurable. It's been a long, rewarding, insightful, fun journey.

About the Authors

Gina C. O'Connor is associate professor of marketing at Rensselaer Polytechnic Institute's Lally School of Management and Technology and academic director of the Radical Innovation Research Program.

She has served as the director of the Lally School's M.B.A. and M.S. programs and associate director of the Severino Center for Technology Entrepreneurship, and she currently serves as director of the executive M.B.A. program at Rensselaer Polytechnic Institute (RPI). O'Connor earned her Ph.D. in marketing and corporate strategy at New York University and her M.B.A. in finance and B.A. in psychology at Saint Louis University. She spent several years with McDonnell Douglas Corporation in contract administration on the AV-8B Harrier program and with Monsanto Chemical Corporation's Department of Social Responsibility.

O'Connor's teaching and research interests lie at the intersection of corporate entrepreneurship and radical innovation, marketing, and commercialization of advanced technology. The majority of her research efforts focus on how firms link advanced technology developments to market opportunities. She has published more than thirty articles in refereed journals and books and is coauthor of *Radical Innovation: How Mature Firms Can Outsmart Upstarts* (2000).

O'Connor won the Lally School teaching award in 2004 and the RPI Alumni Association Teaching Award in 2006. She has consulted with many companies to help them develop management systems for breakthrough innovation.

Richard Leifer is a retired professor of management at the Lally School of Management and Technology, Rensselaer Polytechnic Institute, where he was on the faculty for twenty-three years. He received a Ph.D. in organizational design at the University of Wisconsin at Madison and M.S. and A.B. degrees in psychology and engineering from the University of California at Berkeley.

Leifer taught in the fields of innovation management, organizational behavior, and organizational design. His research focus since 1995 has been on the managerial and organizational aspects of breakthrough or radical innovation. He is coauthor of the book, *Radical Innovation: How Mature Companies Can Outsmart Upstarts* (2000). Leifer has published or presented over seventy refereed articles, and he continues to write and do research. He consults and has speaking engagements for a number of private and public organizations in areas such as innovation management, leadership, change management, organizational design and analysis, and other management issues. He also presents seminars on innovation-related topics on a regular basis at several universities.

Albert S. Paulson is the Frank and Lillian Gilbreth Chaired Professor in the Technologies of Management at Rensselaer Polytechnic Institute. He is the author of over 100 articles, monographs, and books concerning innovation, financial modeling, energy issues, insurance, information extraction, large-scale design, economics, and statistics. Paulson has consulted extensively for national and international firms and governments. He also has served as lead strategist and expert witness in national and international proceedings. His current interests focus on architectural, technological, and financial innovation.

Lois S. Peters is associate professor at Lally School of Management and Technology, Rensselaer Polytechnic Institute (RPI), and a principal investigator at the RPI NSF-sponsored Nanoscale Engineering Research Center. She directed the Lally School's Ph.D. program for ten years. In addition, she is past-president of the International Trade and Finance Association and has been a member of the Institute of Electrical and Electronics Engineers (IEEE) Engineering Management Society (EMS) Board of Governors for ten years. As a member of the EMS Board of Governors and past-vice president of conferences, Peters has organized three international conferences related to the management of technology and innovation and has produced four edited conference proceedings. In 2000, she received an IEEE third millennium medal for outstanding achievement and contribution. She is now chair of the IEEE Technology Management Council case study initiative.

Peters has conducted extensive research on technological innovation, R&D globalization, technological networks including public-private partnerships and commercialization of emerging technologies. A recent stream of research in the latter area includes the role of emotions in the opportunity recognition process. She is a coauthor of *Radical Innovation: How Mature Companies Can Outsmart Upstarts*. Peters has participated in numerous forums focusing on technology policy and management of innovation. She has been a speaker in Japan, the OECD, the European Community, Argentina, Mexico, the U.S., and Thailand among other places. For five months in 1992, Peters was a visiting professor at the Max-Planck-Institute für Gessellschaftsforschung, where she contributed to studies on technological innovation and learned approaches to network analysis.

Peters teaches courses on business implications of emerging technologies; technological entrepreneurship; invention, innovation and entrepreneurship; policy issues in energy and environment; innovation organization and change; and technological change and international competition. Peters has a Ph.D. in biology and environmental health science from New York University.

Index

A

Accelerating business: Analog Devices' migration as, 140*fig*–141; breakthrough innovation uplift for, 147–148, 200; envisioned market entry versus actual entry applications for, 145*t*; growth in sales and inquiries of, 144–146; helping them to gain critical mass, 131; impact on strategic intent, 149; location options and trade-offs available to, 135*t*–136*t*; managing relationship with mainstream organization, 129–131; perceived value of, 149; reduced traceability of, 146–147; spillover to other platforms, 147; structures and processes of, 133–141. *See also* Kodak's System Concept Center (SCC); NBD (new business development)

Acceleration: description of, 20, 117–118; IBM's challenges with, 118–120; importance of, 120–123; Kodak's Systems Concept Center (SCC) application of, 123–128; management systems for, 21*fig*–22; questions for you on, 150. *See also* DNA (discovery, incubation, and acceleration)

Acceleration management system: elements of, 128*fig*; leadership and governance of, 142–144; mandate and responsibilities of, 129–132; metrics and reward systems of, 144–149; resources and skills of, 141–142; structures and processes of, 133–141

Air Products: A-380 airliner of, 122–123; breakthrough innovation capability development at, 1–3; conference attendance by employees of, 59; developing incubator staff at, 105; Growth Board of, 177; internal and external interfaces for BI at, 268; mandates established at, 70; mirrored model approach used by, 173*fig*–174; opportunity articulation at, 63–64; project transitions at, 155

Albany International, 32, 166

Aligned breakthrough innovation (BI), 68–69

Analog Devices, 4, 62, 138–141, 247

Armagnac, D., 144

Austgen, D., 187

B

"Bench of opportunities," 110

Biever, A., 131, 157

Bingham, D., 82–83

Biomax (DuPont), 132

Blaustein, M., 83

Boeing, 32

Breakthrough innovation (BI): aligned, 68–69; categories of, 12t; company focus on, 1; confusing diversification with, 78; culture and leadership that values, 271–273; definitions of, 10–12; distinguishing between R&D and, 259; keys to getting started with, 244–245; looking toward a system for, 260–261; mismatches between senior management expectations and, 198t; motivating the need for, 222–223; open innovation at extreme form of, 162–163fig; organizational function of, 202–203; phases of developing capacity for, 217–228; setting up the infrastructure to support, 236–239; struggle to bring to market, 1–6. See also Innovation management system; Management systems; Opportunities; Organizational capacity

Breakthrough innovation (BI) infrastructure: activities and programs supporting, 237–238; importance of setting up the, 236–237; location and reporting structure as part of, 238–239

Breakthrough innovation (BI) management system: elements of sustainable function of, 266fig; governance at different levels of, 270–271; identifiable organization structure of, 266–267; powerful networks of, 269; rich internal and external interfaces in, 267–269; skills and talent of, 269–270. See also DNA (discovery, incubation, and acceleration)

Breakthrough innovation (BI) phases:1: setting the stage, 218–222; 2: activities associated with setting the stage, 222–227; 3: moving on from the groundwork laid, 228

Breakthrough innovation portfolio: churn rate of, 167–168; cross-portfolio management of, 169–170; diversification of, 165–167; governance at level of, 270–271; incubation, 98–110; managing the health of the, 164–170; pacing the, 169; "portfolio-within-a-portfolio" approach to, 108–109; size of, 164–165

Breakthrough innovation project teams: announcing existence of, 241–242; brokering relationships within, 103; developing incubator staff and, 104–107; exploring resource models to support, 242–244; finding the right people for, 232–236; as incubation talent, 114; maturing breakthrough innovation capability of, 245–257; nurturing of, 103–104; strategic coaching of, 100–102. See also Incubation staff; Innovators

Breakthrough innovation projects: alignment of, 68–69; discovery competency for, 18, 20, 21fig–22, 57–79; governance at the level of, 270–271; handling transitions of, 154–156; incubation at the portfolio level, 98–110; incubation of, 20, 21fig–22, 82–116; knowing when to kill, 256–257; learning plan used for, 88–96; pacing the, 156–157; support function of, 202

Breakthrough innovations capability (BIC): assessing your

organization's capacity for, 23–50; building, 6–10; building blocks of, 20*fig*; defining, 12, 18, 20–22; model of, 19*fig*

C

"Can't get heard" frustration, 159–160*fig*
Capacity influencers: external, 26, 28, 29*t*–30, 40*t*; internal, 28–29*t*, 30–33, 40*t*; source and dynamism of, 29
Carp, D., 125
CEOs: answering to agenda of the, 195–196; orchestrating linkages to senior management and, 196–197. *See also* Leadership and governance; Senior management
Churn rate, 167–168
Coaches/coaching: innovation managers as, 105–107; required throughout incubation period, 113, 114; strategic, 100–102
CommQuest, 5
Communicating innovative objectives, 227
Concannon, M., 118, 119
Connelly, L., 146
Corning Inc.: balancing resources and capability development at, 158; breakthrough innovation capability of, 7–8, 26; commitment to DNA by, 206; discovery competency at, 56–57; EMTG (Exploratory Markets and Technologies Group) at, 56–57; Growth and Strategy Council of, 108, 143–144, 155; innovation model used at, 244; internally based, highly dynamic influencers at, 30; Technology Council of, 155; three R&D spending buckets of, 191
Cotterman, R., 97, 187

Coyne, M., 125
Cryovac, 30
Culture. *See* Organizational culture
Cunningham, P., 118–119

D

Degree of alignment, 67
Dell, M., 78
DeLoso, D., 55
Differential financial incentives, 178, 181
Discovery competency: activities that comprise, 57–58; "can't get heard" frustration and, 159–160*fig*; challenging in building, 77–79; Corning's application of, 56–57; description of, 18, 20, 52–54; foundational knowledge of, 58–60; IBM's application of, 54–56; objective of, 65–66; opportunity articulation of, 62–65; potential business opportunities for, 60–62; questions for you on, 79. *See also* DNA (discovery, incubation, and acceleration)
Discovery management system: alignment of objectives with elements of, 67–71, 72*t*; description of, 21*fig*–22; elements of the innovative, 66*fig*–77; leadership and governance of, 72*t*, 74–76; mandate and responsibilities of, 67–71, 72*t*; metrics and reward systems of, 72*t*, 76–77; resources and skills of, 72*t*, 73–74; structure and processes of, 71, 73
Diversification: as breakthrough innovation capability issue, 219–220; breakthrough innovation compared to, 78; industries case study on, 40–47; NBD (new business development) focus on, 85–86; portfolio, 165–167

DNA (discovery, incubation, and acceleration): avoiding the "can't get heard" frustration, 159–160*fig*; courage to continue with, 159, 160*fig*; emotional roller-coaster of working with, 209–212; executing big ideas, 160–162*fig*; introduction to, 18, 20–22, 48; managing balance of resources and capability development across, 158–164; managing the health of breakthrough innovation portfolio using, 164–170; managing the links and interfaces of, 154–157; orchestrating and incorporating the, 185–213; overview of BI system of, 262*t*–265*t*; questions for you on, 183; relationship of organizational capacity to level of, 204*t*–207; system-level activities of, 151–154. *See also* Acceleration; Breakthrough innovation (BI) management system; Discovery competency; Incubation

DNA management system for innovation: elements of, 170–171; leadership and governance of, 177; mandate and responsibilities of, 171; measuring health of your, 179*e*–181*e*; metrics and reward systems of, 178, 181–182; resources and skills of, 175–177; structure and processes of, 171–175

Donnelly, S., 36, 37, 40, 154, 166

Dow Corning, 52

DuPont: APEX review board of, 108, 177, 249–250; bench inventory of projects practice at, 256; Biomax product innovation of, 132; controlling mandate creep at, 249–250; exploratory marketing groups formed by, 61, 82–83;

Growth Board of, 177; project transitions at, 155; Surlyn product innovation at, 138, 188–190

DuPont Ventures, 83

E

Eastman Kodak. *See* Kodak

Edison, T. A., 35

Einhaus, G., 32, 74, 123–124, 127–128, 169, 200, 233

External capacity influencers: description and sources of, 26, 28, 29*t*; General Electric (GE), 40*t*; highly dynamic, 29–30, 47*t*; low dynamism, 30, 47*t*

Extreme Blue (IBM), 63

F

Food and Drug Administration, 87

"Forced adoption model," 113

Foundational knowledge activities, 58–60

G

GameChangers (Shell Chemicals), 11, 60, 152–153, 165, 243

G.D. Searle, 87

GE Corporate Research, 36–37

GE (General Electric): acceleration resources and skills at, 141–142; ATPs (advanced technology programs) of, 99, 106, 141, 154–155, 165, 166, 174, 176; cross-portfolio management approach at, 170; digital X-ray team initiatives at, 138, 170; influencers of, 30, 40*t*; innovation managers of, 106; legitimizing incubation at portfolio level at, 99–100; mirrored model approach used by, 173*fig*–174; rich breakthrough innovation capacity of, 34–40

Gerstner, L., 4, 221
Giersch, M., 55, 175, 187, 242, 251
Global Technology Center (New York) [GE], 35–36
Google, 32–33, 84
Governance. *See* Leadership and governance
Gustin, C., 125

H

Harreld, B., 8, 9, 165
Henderson, L., 125–126, 128, 130
Hewlett Packard, 127
Hickey, B., 8, 200
Holistic sequential structure, 171–172*fig*
Horn, P., 4, 5
Houghton, J., 30, 206
Hughes Electronics, 4

I

IBM: advisory teams set up by, 109; breakthrough innovation capability of, 3–6; breakthrough opportunities at, 8; challenges with developing the silicon germanium business, 118–120; discovery competency at, 54–56; Extreme Blue program at, 63; innovation embedded into culture of, 221; multialigned opportunities designated at, 71; 1 (H1) business label used by, 8; self-similar model used at, 172–173*fig*; strategic coaching used at, 101; 3 (H3) business label used by, 8, 55
IBM's EBOs (emerging business opportunities) program: announcements introducing, 242; controlling the process at, 251; discovery competency at, 54–56; incubation in, 83–84; incubator

staff development at, 104; new service opportunities launched by, 247; portfolio growth of, 165; process of declaring new, 269; strategic coaching at, 101; structure and processes of, 112; traceability at, 146–147
Immelt, J. R., 30, 35, 36, 38, 176, 242
Incubation: comparing new product marketing activities to, 86*t*–87*t*; definition of, 83–87; individual opportunities for, 87–97; introduction to, 20, 82–83; management systems for, 21*fig*–22; at the portfolio level, 98–110; questions for you on, 116. *See also* DNA (discovery, incubation, and acceleration)
Incubation management systems: elements of, 111*fig*; leadership and governance of, 114–115; mandate and responsibilities of, 111; metrics and reward systems of, 115–116; resources and skills for, 113–114; structure and processes of, 112–113
Incubation portfolio: described, 98; developing staff for, 104–107; legitimizing, 98–100; monitoring, 107–110; providing project team support at, 100–104
Incubation staff: development of, 104–107; incubation talent of, 114. *See also* Breakthrough innovation project teams
Incubation talent: recognizing and hiring, 104–107; two levels of, 114
Infrastructure: activities and programs supporting, 237–238; importance of setting up the, 236–237; location and reporting structure as part of, 238–239

Innovation function: appropriate metrics and rewards supporting, 271; BI processes and tools supporting, 270; BI skills and talent supporting, 269–270; culture and leadership valuing BI and, 271–273; DNA management system for, 171–182; governance supporting, 270–271; identifiable organization structure required for, 266–267; management system supporting sustainable, 260–261, 266fig; powerful networks supporting, 269; rich internal and external interfaces supporting, 267–269

Innovation management system: importance of creating a, 215–217; phases of developing capacity for, 217–228. See also Breakthrough innovation (BI); Management systems

"Innovation principles," 127–128

Innovators: demanding personalities of, 232–234; as innovation process facilitators, 201–202; lack of career path available to, 235; organizational role of, 251–252; process anxiety by, 250–252; undertraining available to, 235–236. See also Breakthrough innovation project teams; Orchestrators

Intel, 130, 131, 157

Internal capacity influencers: description and sources of, 28–29t; General Electric (GE), 30, 40t; highly dynamic, 30–31, 47t; low dynamism, 31–33, 47t

ISO 9000, 14

J

James, P., 44–45

John F. Welch Technology Center (India) [GE], 36

Johnson & Johnson Consumer Products, 100, 165–166, 205, 242

Junkins, J., 28, 33

K

Kodak: advisory teams set up by, 109; aligned breakthroughs at, 67–68; business accelerator approach of, 123–128; holistic and sequential structure of, 171–172fig; mandates followed at, 70–71; NEXT team of, 125, 126, 143

Kodak's System Concept Center (SCC): as business accelerator, 123–128; continued innovation function started by, 200; as innovation hub, 8–9, 70–71; monitoring mandate creep at, 192; opportunity generation by, 60; project-level skills and resources allocated to, 96–97; structure and team members of, 108, 233–234. See also Accelerating business

L

Leadership and governance: of acceleration management system, 142–144; of breakthrough innovation management system, 270–271; breakthrough innovation valued by, 271–273; of discovery management system, 72t, 74–76; of DNA management system, 177; of incubation management system, 114–115; incubation portfolio, 107–110; management system, 15; maturing BI capability and role of, 252–256; orchestrating linkages to company, 196–198t. See also CEOs; Senior management

Learning loop, 95, 110

Learning Plan Template: assumptions underlying, 94; steps and process of using, 90, 91e–92e, 93

Learning plans: effects of, 94–96; four uncertainties recognized by, 88–89; learning loop approach to, 95; template for, 90–94
Legitimizing incubation, 98–100
Leonardo da Vinci, 58
Loose, J., 30
Lucent, 125, 178

M

Madison, G., 40, 41–42, 45–46, 47
Management systems: acceleration, 128fig–150; for breakthrough innovation, 261, 266fig–273; comparison of mainstream and innovation, 17t; counterbalancing imbalances of, 163–164; definition and elements of, 13fig; description of successful, 16–18; discovery, 66fig–77; incubation, 111–116; leadership and governance element of, 15; mandate and responsibilities element of, 14; metrics and rewards systems element of, 15–16; resources and skills element fueling, 15; structure and processes element of, 14. *See also* Breakthrough innovation (BI); Innovation management system
Managing inward, 254–255
Managing outward, 253–254
Managing upward, 252
Mandate and responsibilities: of acceleration management system, 129–132; clarifying the innovation function, 223–225; of discovery management system, 67–71; of DNA management system, 171; of incubation management systems, 111; management systems, 14; monitoring mandate creep, 191–193

Mandate creep: definition of, 70; maturing BI capability and, 249–250; monitoring, 191–193
Market uncertainties, 88–89
Martino, M., 125
Matheson, N., 205, 242
Maturing BI capability: characteristics of, 247fig; idea generation and scope, 246–248; killing projects and, 256; leadership demands during process of, 252–256; levels of metrics for, 257, 258t; mandate creep and, 249–250; overview of, 245–246; process anxiety and, 250–252; system resource level, timing, and balance elements of, 256
Mayer, M., 84
McElhenny, T., 37
MeadWestvaco, 60, 105, 131, 144, 165
Metrics and reward systems: of acceleration management system, 144–149; breakthrough innovation and levels of, 257, 258t; of breakthrough innovation management system, 271; of discovery management system, 72t, 76–77; of DNA management system, 178, 181–182; of incubation management system, 115–116; management system, 15–16
Meyerson, B., 3–6, 118, 119
Miller, J., 56
Mir, J., 123–124, 128
Mirrored model, 173fig–174
Mitch's story, 235
Modern Plastics, 189
Moghadam, O., 130
Moonshots, 11
Motivation: for initiating BI capability, 230t; need for breakthrough innovation (BI), 222–223
Motorola, 125

N

NBD (new business development): diversification focus of, 85–86; Kodak's use of, 125; shortage of skills in creating, 234–235. *See also* Accelerating business

Nelson, M., 65

Newhouse, M., 57, 95, 187

Nintendo, 138

Nintendo Wii, 138–139

Nortel Networks, 109, 181

Northern Telecom, 4

Nurturing project teams, 103–104

NutraSweet, 87

O

Open innovation at extreme, 162–163*fig*

Opportunities: "bench of opportunities," 110; discovery competency and potential, 60–62; discovery competency articulation of, 62–65; "forced adoption model" for, 113; incubation and individual, 87–97. *See also* Breakthrough innovation (BI)

Opportunity activities: discovery competency articulation, 62–65; discovery competency generation, 60–62; project versus platform approach to generate, 61–62

Orchestration activities: challenges of pioneering, 207–212; considering capacity and DNA capacity levels simultaneously, 204*t*–207; definition of the, 186; developing an innovation strategy, 191*fig*; under munificent capacity, 203; orchestrator's agenda for, 187–201; questions for you on, 213; under stressed capacity, 201–203

Orchestration challenges: emotional roller-coaster of DNA system, 209–212; working with innovators, 207–209

Orchestrators: agenda of, 187–201; challenges facing, 207–212; examples of, 187; questions for you on, 213; roles and influence of, 186–187. *See also* Innovators

Orchestrator's agenda: managing the BI function's perceived role, 188–191*fig*; monitoring mandate creep, 191–193; orchestrating linkages to company leadership, 196–198*t*; orchestrating linkages to other corporate functions, 190; orchestrating to get things done, 194–196; orchestrating transitions of project to operating units, 193–194; orchestrating uplift of the innovation function, 200

Organizational capacity: assessing your company's, 48–50; conducting an audit of the, 225–227; constrained, 40–47*t*; definition of, 26; external influencers on, 26, 28, 29*t*–30; internal factors of, 28–29*t*, 30–33; lessons on, 47–48; maturing breakthrough innovation, 245–257; model of breakthrough innovation, 27*fig*; munificent, 34–40*t*; relationship of DNA level to, 204*t*–207; summarizing, 33–34; tracing example of birth and death of, 23–25. *See also* Breakthrough innovation (BI)

Organizational culture: affecting approaches to discovery, 74–75; breakthrough innovation valued by, 271–273; building an innovation, 229, 231–232; ideas stimulated by changes to, 79

Organizations: breakthrough innovation education of mainstream,
131–132; building capability for
breakthrough innovations, 6–10;
focus on breakthrough innovations
by, 1; importance of acceleration
capabilities to, 120–123; relationship between accelerating business
and mainstream, 129–131; uncertainties faced by, 88–89, 122–123

P

Pierantozzi, R., 59, 70, 105, 155, 187
Pingle, L., 60
Portfolio. *See* Breakthrough
innovation portfolio
Processes. *See* Structure and processes
Projects. *See* Breakthrough
innovation projects

R

R&D (research & development),
259
Radical innovations, 11
RAM's "click test," 189
Rees, R. W., 188–189
Resource uncertainties, 89
Resources and skills: for acceleration
management system, 141–142;
balancing development capability
with, 158–163; for breakthrough
innovation management system,
269–270; for discovery management system, 72*t*, 73–74; of DNA
management system, 175–177; for
incubation management system,
113–114; management system, 15
Reward systems. *See* Metrics and
reward systems
Reward/punishment approach, 178,
181–182
Royal Dutch Shell, 152, 153

S

SCC (System Concept Center). *See*
Kodak's System Concept Center
(SCC)
Scope change innovations, 11
Sealed Air Corporation: acceleration
as defined by, 129; Business Innovation Board of, 177; continued
innovation function at, 200; innovation opportunities nurtured
at, 8; market participation
approach taken by, 96; merger
between Cryovac and, 30–31;
opportunity generation at, 60–61;
Technology Identification Process
(TIP) team of, 97
Self-similar model, 172–173*fig*
Senior management: BI group leader's
upward management toward, 252;
mismatches between BI realities
and expectations of, 198*t*; orches-
trating linkages to senior manage-
ment, 196–197. *See also* CEOs;
Leadership and governance
September 11, 2001, 26
Shell Chemicals (GameChangers),
11, 60, 105, 152–153, 165,
243, 268
Shell Exploration and Petroleum
(Shell E&P), 152, 153
Six Sigma approach, 112
Skills. *See* Resources and skills
Sony PlayStation, 138
Sousa, N., 96–97, 124–125
Spedden, R., 60, 187
Staff. *See* Breakthrough innovation
project teams
Stoeffel, J., 125
Strategic coaching, 100–102
Structure and processes: of accelera-
tion management system,
133–141; breakthrough

Structure and processes (*continued*):
innovation tools and, 270; of
discovery management system, 71,
73; of DNA management system,
171–175; of incubation management system, 112–113;
management system, 14
Surlyn story (DuPont), 138,
188–190

T

Technical uncertainties, 88
Technology Center (China)
[GE], 36
Technology Center (Germany)
[GE], 36
Texas Instruments (TI), 28, 33
Thomas, L., 63–65, 70
3M, 61, 65, 75, 112, 200, 244
Time horizon issue, 166

U

Uncertainties: associated with acceleration, 122–123; learning plan
recognition of, 88–89; market,
88–89; organizational, 89;
resource, 89; technical, 88
Uplift: accelerating business,
147–148; orchestrating innovation, 200
U.S. Food and Drug Administration, 87

V

Vanrose, J., 45

W

Welch, J., 30, 34, 35
Wolpert, J., 63

Z

Zank, G., 52